The River Doesn't Wait

A Senior Executive Guide to Navigating the Surge of Agentic AI and Pulling Away From Your Competitors

Blaine Mathieu

Murphys Business Press

Published by Murphys Business Press
Murphys, California

Paperback ISBN: 979-8-9966246-0-7
Hardcover ISBN: 979-8-9966246-1-4
Ebook ISBN: 979-8-9966246-2-1

First Edition

Contents

Dedication

For my wife Wendy Mathieu, who has not only been my lifelong love and friend but also a business partner and confidant who has helped me work through the thought process around much of this book as it slowly crystallized over the last 30+ years, and accelerated over the last two.

For my son Aiden and his wife Jing, who are young adults and digital natives, and are now bravely working through how to succeed inside this massive wave of transformational change now upon us.

Finally, for my new grandson, Marc. Decisions we make and paths we take today, whether in business or in society at large, will have a profound effect on the world in which he grows up. Let's make the right choices.

Where Senior Leaders Live

Foreword

Every decade or so a technology arrives that enterprises adopt faster than they understand. Agentic AI is the current instance. Systems that once answered questions now take actions, and they take them continuously, at machine speed, across workflows that no single person watches end to end. The capability is real. What most organizations have not yet reckoned with is that capability of this kind exposes a weakness that was always present in their operations but never before consequential: the gap between what their systems are supposed to be doing and what those systems are actually doing.

I have spent my career on the architecture of that gap. In the research my colleagues and I conduct at the Mindful AI Foundation, we have a name for it. We call it coherence debt: the accumulating mismatch between the purposes, policies, and constraints an organization believes govern its systems, and the behavior those systems actually produce under real conditions. Coherence debt is not a software bug. It is structural. It grows quietly whenever knowledge, memory, and governance are treated as things you bolt on around a system rather than build into it. And it stays hidden, because at human speed the gap is small enough to absorb. A few errors a week. A policy honored in most cases. Manageable.

Agentic AI removes the thing that kept it manageable, which was slowness. When systems act continuously and autonomously, a gap that produced a handful of quiet errors a week can produce them by the thousands before anyone notices a rule was broken. The debt does not change in kind. It changes in scale, and scale is what turns a tolerable inefficiency into a structural risk. This is why I think the executives now asking why their AI investments are not paying off are, in many cases, asking the wrong question. The problem is rarely the model. The problem is that a powerful new engine has been bolted onto an operating logic that was never designed to be coherent at this speed.

That diagnosis is where my work lives. It is not, however, where the solution lives for the person reading this book. From the systems level, coherence debt looks like an architectural problem, and there is an architectural answer to it. That answer is a longer story, and it is told elsewhere. Because the moment you move from the engineering of these systems to the running of an enterprise that depends on them, the problem changes character. It stops being a question of how the machine is built and becomes a question of how the organization is designed. Who decides what an agent is allowed to do. What it is permitted to see. Where authority sits when a human is no longer in the loop on every decision. Those are not technical questions. They are leadership questions, and no technology officer can answer them on a leader's behalf.

This is the level at which Blaine works, and it is why I am glad he wrote this book rather than waiting for someone to write it for him. He is a Fellow of the Foundation whose work on the operator's side of these questions I have followed closely, and what I value in it is precisely what is hardest to find: a grasp of what these systems genuinely do, paired with a refusal to let the conversation stay technical when the decisions that matter are not. The book skips the vocabulary of models and tokens and writes where senior leaders actually live, in the architecture of decisions, authority, and organizational design. Its central distinction, between retrofitting AI onto your existing operating logic and reimagining that logic with agents as participants in the work, is the right frame for the choice every leader is now facing, whether they have named it or not.

I will offer one closing thought, because it is the one I most want a leader to carry into these pages. You cannot delegate this. Your engineers can build the systems. Only you can set the tolerances, define the boundaries, and accept accountability for how your enterprise behaves when it is moving faster than you can supervise it. That responsibility does not transfer. This book is the clearest account I know of what it requires.

Dr. Rao Mikkilineni, PhD

Founder, Mindful AI Foundation

Chapter One

The Curve Is Real

Where the technology stands, and why the gap is widening

Two years ago, the question senior executives asked about AI was when. When will we start. When will the technology be ready. When will we have to take it seriously.

That question is mostly gone. By the spring of 2026, the question I hear from senior executives has shifted. They are asking, with varying degrees of frustration, why isn't this working yet?

Why is the productivity gain not showing up in the P&L? Why have we canceled a third of our pilots? Why does our AI strategy slide deck look identical to our competitor's, and theirs looks identical to ours, and neither of us is convinced any of it actually matters? Why are we reading about JPMorgan and Moderna and BBVA, and our own organization still seems to be running roughly the same way it did in 2023?

The answer is uncomfortable. Most enterprises are doing this wrong, in the same wrong way, for the same wrong reasons. The handful that are doing it right are pulling away faster than the rest can see.

That is the situation this book will describe.

• • •

From a practical perspective, this book is about agentic AI, its potential impact on your organization, and how you should think about it. I wrote it primarily for senior business executives in positions of power at large enterprises. Likely VP or above. Owner of a function, a P&L, a region, or a major program, or possibly even the CEO. The kind of leader who has navigated at least one major digital transformation already, who isn't a technologist and isn't trying to become one, but whose board is asking about AI strategy, and whose CIO or CTO is presenting incredible new technologies to them. The kind who has read enough to know that the wave of agentic AI matters and is now trying to figure out what to do about it.

If that is you, this book is for you. Your McKinsey advisors might find it useful too, and may also find it competing with their decks.

If you have made it this far into a stack of agentic-AI newsletters, podcasts, videos, books, and conferences, you have my respect. By the time you picked this one up, you had already worked your way through the McKinsey playbook, the BCG framework, the Bain perspective, the Deloitte guide, the Accenture point of view, the many Microsoft white papers, the OpenAI enterprise brief, the IBM analysis, and at least one chapter-and-verse PDF that came out of Davos. Your inbox is a graveyard of webinar invitations. Your assistant has stopped forwarding the briefings.

I am not going to pretend any of those analyses are bad. (As a former Gartner analyst myself, I used to be responsible for some of them.) Most of them have something useful in them. Cumulatively, though, they have a problem. They were produced by people whose business model depends on a particular conclusion. The consulting firms need you to need consultants. The platform vendors need you to buy platforms. The think tanks need you to fund research. None of these incentives are evil. They just bend each frame.

What you have not had is an operator's read. A sober synthesis from someone whose payoff function is being right about how this plays out, not selling you a service. (As of the time of this writing, I was not employed or paid by any company in the agentic AI space.) That is the deal I am making with you, and it has consequences for what is in this book.

What this book is, and isn't

The deal goes both ways. You are not getting a primer on AI. You will not learn how a transformer works, or what fine-tuning means, or how to evaluate a model on a benchmark. There is plenty of material on those topics, much of it free, written by people who know the technology better than I do. Your CIO already has it bookmarked.

You are also not getting a vendor scorecard. The platforms are still consolidating, and any list I give you would be out of date by the time the book hits print. If you want to know whether to buy from Microsoft or Anthropic or OpenAI or Google or one of the dozen more focused challengers behind them, you will find better signal from your CIO than from a book.

What you are getting is a synthesis you can finish on a cross-country flight. Strategic clarity instead of technical depth. The shape of the choice you are already making, whether you have noticed it or not, and the way to make that choice deliberately rather than drift into one of its bad endings by default.

There is one more scope decision I owe you up front. This book is about how you run your organization. It is not about how your organization shows up to its market.

By "how you run it" I mean the internal operating logic of the enterprise. Processes. Workflows. Organizational structure. Roles. Who has authority over what. How decisions get made. How context flows. How autonomy gets granted, and how governance keeps it from going off the rails. The interior of the machine.

By "how it shows up to its market" I mean things that genuinely matter and are genuinely outside the scope of this book. Agentic commerce, where buyers' agents will eventually transact with sellers' agents at machine speed. Customer-interface redesign, where the chatbot replaces, or fails to replace, the rep, the dealer, the branch. Pricing-model shifts, from seat-based to outcome-based to whatever comes after that. Brand-mediated versus agent-mediated customer experience. Whether your products themselves should become agents.

These are real questions. They will reshape your industry. Some of them may end up in a future book but they are not in this one.

This one is the more boring half, and the half that has to come first, because before you can ask whether your customer-facing front end becomes an agent, you have to know whether the rest of your enterprise can plausibly be operated alongside one.

If you came looking for the agentic-commerce chapter, this is your moment to pass this book on to a colleague.

For everyone still here, that scope decision is also why I think this book has a chance to be useful. The market-facing material has been written about more than the operational material, partly because the market-facing material is sexier, partly because the consulting firms got to it first, partly because the operational material is harder to write about without sounding either like a primer or like a slide deck. Most of what gets published on agentic AI either overgeneralizes ("AI will transform everything") or overspecializes ("here is how to fine-tune a customer-service bot"). The middle altitude, where the senior executive actually lives, has been underserved.

That middle altitude is where I am trying to live.

Why now

There is a version of the why-now argument I cannot quite make. It would go like this: there is a moment, possibly right now, beyond which it will be too late to change course. After this window closes, the laggards will be permanently locked out. So move now, before the window slams.

I cannot make that argument because I don't believe it. The technology curve is not going to close. Foundation models are going to keep getting more capable. Agents are going to keep getting more useful. A slow-moving enterprise that wakes up in 2030 will still find willing vendors, plentiful talent, and a mature ecosystem ready to onboard them. What it won't find is a market dynamic that has waited for it.

It is not a technology window that is closing. It is a competitive gap, and it is widening.

Two or three years ago, a senior executive could reasonably stand pat on AI. The technology was hyped, but the value capture was thin. The handful of companies running ahead were small, mostly digital-native, and easy to dismiss as edge cases. Standing still cost very little.

That is no longer true. By the spring of 2026, the leaders running ahead are not just digital natives. They are some of the most conservative institutions in their industries. JPMorgan Chase, the largest US bank by assets, reports more than 450 AI use cases in production by April 2026, a roughly $19.8 billion technology budget, and documented value capture in the $1.5 billion to $2 billion range. BBVA went from 3,300 employees with internal AI access in May 2024 to 11,000 by mid-2025, and by December 2025 had announced an expansion to all 120,000 employees globally. Genpact, an enterprise services company most readers' boards have probably heard mentioned in passing, is publicly describing what its CEO calls "Agentic Operations" as a new operating model on its earnings calls.

Meanwhile, in the same window, roughly 95 percent of enterprise AI pilots produced no measurable P&L impact, by MIT NANDA's State of AI in Business 2025 count. Gartner's 2025 forecast was that more than 40 percent of agentic AI projects would be canceled by the end of 2027.

Read those numbers together. The leaders are running hundreds of use cases at industrial scale and capturing measurable value. The middle 95 percent are running pilots that don't go anywhere, with the worst of them about to be canceled in the next eighteen months.

That gap is the window. It is widening because the leaders compound.

The Boston Consulting Group titled its 2025 report on AI value capture "The Widening Gap." McKinsey's State of AI segmentation puts the high-performer cohort at roughly 6 percent of organizations, with another 39 percent experimenting and 23 percent scaling agents in at least one business function, though narrowly. Microsoft's Agent Readiness Survey, published in September 2025, scores companies on strategy and execution and finds that the most-ready firms, its 'Achievers,' expect to scale agent deployments roughly two and a half times faster than the least-ready, its 'Discoverers.' MIT NANDA, in its widely-read Generative AI Divide paper, describes the gap between frontier and median enterprise behavior as appearing to be widening.

A reasonable executive read of this evidence: the cost of inaction in 2024 was low. The cost of inaction in 2027 will be visible on quarterly calls. By 2029, the cost of inaction may be existential for some firms in industries we cannot yet name.

Jamie Dimon, on a World Economic Forum 2026 Davos panel, put it the way only Dimon can.

"If you put your head in the sand, you will lose."

Dimon is right. The more interesting voice, though, is your nearest competitor's CEO, whose AI strategy slide deck looks suspiciously like yours, and whose results so far look about like yours. The interesting question is whether their next strategy slide deck still looks like yours.

This book exists because the noise around the window is so deafening that thoughtful executives cannot hear themselves think. By April 2026, every major firm with an opinion has published its agentic-AI manifesto. Most of them disagree with each other. Many of them are designed to sell you something. Senior leaders looking for a clear-eyed read of where to plant their feet are wading through frame after frame, each of them partial, none of them the whole picture.

That is where I am trying to land. Not a new framework. A synthesis of what is actually happening.

What I mean by agentic

A note on what I mean by "agentic," because the term has been used to mean almost everything AI by now and is therefore well on its way to meaning almost nothing. (Remember "Digital Transformation"?)

For the purposes of this book, agentic AI refers to systems that exhibit five characteristics together.

1. They pursue goals with some degree of autonomy, deciding next steps rather than waiting for instructions on every action.

2. They use tools, interface with and invoke other software, query databases, manipulate files, and act outside their own conversational frame.

3. They reason across multiple steps rather than producing single-shot outputs, planning, executing, observing, and adjusting.

4. They persist, holding context across time and increasingly across days or weeks.

5. And they operate at machine speed and continuous duty cycle, not waiting for office hours and not getting tired at five in the afternoon.

That definition deliberately excludes a lot of things sometimes filed under "agentic." A chatbot answering customer questions in a single turn is not agentic. A spreadsheet that runs a regression is not agentic. A Robotic Process Automation (RPA) bot that follows a fixed script is not agentic; what makes RPA RPA is precisely the absence of the first three properties. Generative AI that drafts a memo for human review is not agentic, even though much of the underlying technology is shared.

What is agentic, by this definition, is the system that takes my instruction to "find me three good investor candidates for our Series B," browses the web, reads ten dozen LinkedIn profiles, drafts an outreach email per candidate, schedules them on my calendar, and follows up if they don't respond within four days. It is the system that reconciles a vendor's invoice against my purchase order, flags the discrepancies, escalates the ones over a threshold, and quietly resolves the ones below it. It is, more importantly, the system that does either of those things in your enterprise alongside thirty thousand other agents doing related and unrelated work, all of them coordinating.

That is the technology this book is about. Not the chatbot. Not the autocomplete. The thing that does work, autonomously, alongside your people.

The trajectory

A year ago, when I told people I was building my own AI agent, the most common response was a polite version of "good luck with that." Today, when I tell people I personally run several of them, the most common response is "what platform are you on?"

The technology has moved.

I don't want to overstate this. The systems are not magic and they are still not simple. They make mistakes. They confidently produce wrong answers. They get distracted. They sometimes have to be talked back from the ledge. Anyone working with them daily can tell you stories. (How different are these from the stories we tell about our human workers?)

What is true, though, is that the rate at which they are getting more capable is the most striking trajectory I have seen in three decades of watching the AI field. And it is striking in a way that is becoming hard to dismiss.

The sharpest measurement of that trajectory I have seen comes from METR, an independent evaluation lab. METR measures agent capability by a deceptively simple proxy: the length of a human task an AI agent can complete with at least 50 percent reliability. In 2019, that number was on the order of seconds. By 2024, it was on the order of an hour. By the time you are reading this, it will likely be on the order of a working day. The doubling time on this measurement has held at roughly seven months for several years now; in the most recent year, it has tightened, by some readings, to about four months.

Here is why that matters. Most exponential trajectories in technology are exponential in something the executive doesn't directly care about. Compute density. Transistor count. Parameter scale. Useful, but a layer removed from operational reality. METR's measurement is, by design, in units the executive does care about. How much of an actual employee's day of work can the system handle? Not how big the model is. How much actual work.

If the answer two years ago was "seconds of work," and the answer last year was "an hour of work," and the answer this year is "most of a day," then the operational implications shift roughly as fast.

That is the curve. Now the part most vendor decks won't show you.

The same lab that produced that time-horizon trajectory also ran a randomized controlled trial last year on experienced software developers using AI tools on familiar codebases. The result, which METR publicly published, was that the developers in the study were roughly 19 percent slower with the AI tools than without them.

I happen to think that is the most useful single result in the entire research base. Two reasons.

First, it tells you something true: capability climbing on a benchmark is not the same thing as productivity gain in real work. The two things are related, but the path between them is full of friction. Skilled humans on familiar work, working in tools they already know, are sometimes slowed down rather than sped up by the assistant. The technology has to integrate. People have to adapt. Workflows have to change. None of that happens instantly.

Carnegie Mellon researchers, separately, have built a benchmark called TheAgentCompany that simulates a realistic multi-step office workflow and measures how often an agent can complete it autonomously. As of this writing, even the leading agents on the public leaderboard fail the majority of tasks, which are designed to be representative of office knowledge work.

That is what the ceiling looks like at the time this book was written.

So which is it? Capability is rocketing toward an inflection point, or capability is still well below "agents run themselves." The answer is both. The trajectory is genuine. The current ceiling is also genuine. What is happening, and what makes the moment interesting, is that the trajectory is moving the ceiling.

In the second week of April 2026, two frontier labs released models that pushed the ceiling visibly higher. Both shipped at one-million-token context windows (think of it as the size of a model's short-term memory). Both were positioned, in their public framing, as agent runtimes, infrastructure for autonomous work. On OSWorld, a benchmark designed to measure agentic computer use on tasks an office worker would recognize, scores roughly doubled in the eleven months from May 2025 to April 2026.

These numbers aren't here to impress. They aren't here to scare. They're here to set up the question:

if the trajectory keeps going, what does the curve look like by the time this book is two years old?

I do not know. Neither does anyone else, and you should be deeply suspicious of any author or analyst who claims they do. What I can tell you is that the rate is unpredictable, the direction is not, and the inflection points are still arriving faster than most enterprises can absorb them.

• • •

That is the lab evidence. Now the field evidence.

JPMorgan Chase, the largest bank in the United States by assets, is running over 450 AI use cases in production as of April 2026, with at least 200,000 employees actively using its internal LLM Suite platform. The figure that should hold your attention is neither of those numbers. It is that they exist at all, at JPMorgan, an institution that does not take experimental risk lightly. JPMorgan is not Klarna. JPMorgan is not a Silicon Valley scale-up with a YOLO mentality and a wall of beanbag chairs. JPMorgan has more compliance and risk officers than most countries have lawyers. And JPMorgan is running 450 use cases and reinventing core processes around them.

In April 2024, Jamie Dimon's annual shareholder letter framed AI as transformational, comparable to "the printing press, the steam engine, electricity, computing and the Internet." That was the augmentation framing. Help every employee do their existing job a bit better. Two years later, in his April 6, 2026 letter, the framing has changed: "AI will affect virtually every function, application, and process in the company." The shift from "augment virtually every job" to "affect virtually every function, application, and process" is not a wording flourish. It is a different thesis about what is happening.

A senior JPMorgan executive, speaking in April 2026 about the firm's new agentic KYC (Know Your Customer) system going to production, was even more direct. Describing the firm's client onboarding process, the executive said it was undergoing "complete reimagination, which is agentic-first. That means agents will orchestrate the onboarding of clients in the workflow."

I want to point at a specific phrase there. Complete reimagination. Agentic-first. Those are the words of an executive describing a new operating logic. We will come back to that vocabulary in the next chapter, because it is the vocabulary the rest of this book is built on, and it is striking that the most important voices using it are not coming from the AI labs or the consulting firms. They are coming from the firms doing the work.

JPMorgan is the load-bearing example because of its scale, its industry, and its conservatism. But it is not alone. Doug McMillon, then-CEO of Walmart, speaking at Harvard Business Review's Future of Business event in November 2025, put it this way: "Every job we've got is going to change in some way - whether it's getting the shopping carts off the parking lot, or the way our technologists work, or certainly the way leadership roles change." He added, helpfully for the executive who is worried this book is going to tell them to fire their workforce, that the goal is to "equip everybody to be able to make the most of the new tool s… and still be a really large employer years from now." Both are consistent with the position McMillon is taking, which is neither retrofit-only nor full-reimagine (two terms the next chapter will pin down). We will come back to that, too.

Genpact's BK Kalra, on the firm's Q4 2025 earnings call in February 2026, described the company's go-forward operating model as "Agentic Operations" and told investors: "We are fundamentally reshaping how businesses operate, and we are doing so at speed." Whatever you think of Genpact's particular execution, the language matters. New operating model is how executives talk about a structural shift.

If you read enough earnings call transcripts, the pattern is consistent. The leaders are describing AI as a change in the substrate.

• • •

Key takeaways

- **The technology curve is real, measurable, and accelerating.** It will not wait for overly cautious adopters.
- **The leaders pulling ahead include some of the most conservative institutions in their industries.** Not the disruptors you would expect.
- **Agentic AI means systems doing autonomous work, not better assistants.** They plan, execute, and adjust without a human at each step.
- **The gap between top performers and everyone else is widening.** Multiple independent data sources confirm the divergence, and it is not slowing.
- **The cost of inaction was negligible in 2024.** By 2027, it will show up on quarterly calls.

• • •

That is where the trajectory has gotten to. The case that this is real is done. The next question, what are you going to do about it, turns out to have only two real answers. Most executives are choosing one of them without realizing they are choosing.

Further reading

Companion website: riverdoesntwait.com

A handful of readings worth your time if you want to go deeper on the material in this chapter. All sources below have live links on the book's companion website, kept up to date as URLs change. None of them is a substitute for finishing this book first; all of them are worth a few hours afterward. Sources surfaced here will not be repeated in subsequent chapters' Further Reading sections, even where the source is referenced again later.

METR Time Horizon trajectory

The capability-trajectory measurement the chapter's "the curve is real" section anchors to. The METR site lets you toggle between log and linear scale; the linear view is the one that captures the current's acceleration most viscerally. Worth visiting periodically rather than reading once. The chart updates as new models land. METR also publishes the randomized controlled trial showing experienced developers were slower with AI tools than without them, which is the result that earned the chapter's trust in this source (but may also be out of date by the time you read this).

Carnegie Mellon, TheAgentCompany benchmark

The honest counterweight to capability-trajectory enthusiasm. TheAgentCompany simulates a realistic multi-step office workflow and measures how often an agent can complete it autonomously. The leaderboard is public and updates over time. As of this writing, the leading entries on the leaderboard still fail a slim majority of tasks. That is the ceiling worth holding in your head alongside the trajectory. Both readings are true at the same time.

Jamie Dimon's annual shareholder letters, JPMorgan Chase

The trajectory of one institution's thinking about AI, told through the chairman's annual voice across roughly a decade, with agents entering the story only in the most recent letters. The chapter quotes the April 2024 letter ("the printing press, the steam engine, electricity, computing and the Internet") and the April 2026 letter ("AI will affect virtually every function, application, and process in the company"). Read the two in sequence and the rightward shift is unmistakable. Earlier letters in the archive (2017 onward) show the start of the arc. JPMorgan's investor relations site holds the full set; the secondary trade-press coverage is reliable for headlines and unreliable for context.

JPMorgan's agentic KYC redesign, April 2026

The "complete reimagination, agentic-first" framing of client onboarding, attributed to a senior JPMorgan executive in coverage of the firm's new agentic KYC system going to production at the end of April 2026. The chapter relies on this framing for the move from "AI as augmentation" to "AI as operating logic." What is striking is the operational specificity: the same redesign that produced the language is also the redesign that took the firm's KYC cycle from a multi-day process to under a minute. That is what reimagine looks like in operating terms.

Boston Consulting Group, "Are You Generating Value from AI? The Widening Gap" (2025)

The clearest articulation of the leader-laggard divergence argument from a major consulting firm. The title alone signals the thesis. Read it for the gap data and the diagnosis. The prescription is shaped by what the firm sells, as any firm's would be, which colors the framing without undermining the data behind it.

MIT NANDA, "The GenAI Divide" / "State of AI in Business 2025"

The 95-percent-no-P&L finding lives here, along with a more transparent methodology section than most things in the research base. Aditya Challapally is the lead author. The paper is also the source of several of the surprising-detail findings that surface throughout this book, including the mid-market-outperforms-large-enterprise data and the shadow-AI-outperforming-sanctioned-AI observation. Worth reading in full rather than reading about.

Chapter Two

The Choice You're Already Making

Why agentic AI requires every executive to decide what kind of company they intend to run

Two paths

Here is the choice. I will state it cleanly first, then unpack it.

You can use agentic AI to make your existing operating logic run faster, cheaper, or at higher volume. Or you can use it to enable a redesign of the operating logic itself.

Most enterprises, when they talk about "deploying AI," mean the first thing. The second thing is what JPMorgan meant by "complete reimagination, agentic-first." The first thing is sometimes called retrofit. The second thing is sometimes called reimagine. The labels matter less than the fact that they are different choices, that every enterprise is implicitly making one of them, and that very few enterprises have made the choice deliberately.

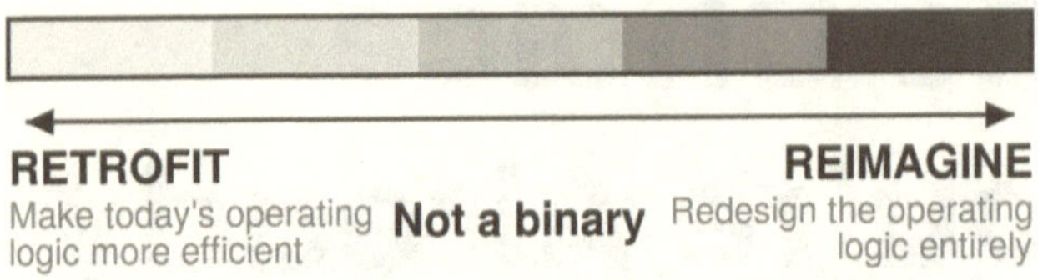

The strategic retrofit-to-reimagine spectrum

By "operating logic" I mean something specific. I mean the invisible architecture that governs how your enterprise actually works. How authority flows. How decisions get made and by whom. How work moves across functions. Who checks whom, and how many layers of checking there are. How accountability is assigned when something goes wrong. How information travels from where it is collected to where it is used. How handoffs happen between teams.

Almost none of this is written down in any one place. It is encoded across your org chart, your approval workflows, your meeting cadences, your management hierarchy, your compensation structure, your performance review system, and a thousand small habits no one has ever named. Most of it is so deeply assumed that the people operating inside it cannot see it. It is simply how work gets done here.

Operating logic is the substrate. Process is what runs on top of it.

When most enterprises automate or improve a process today, they are working on top of the substrate. They are speeding up an approval workflow, but the workflow still has the same number of approval layers. They are reducing cycle time on customer onboarding, but the same set of teams still own the same handoffs. They are taking 90 minutes of work down to 30 minutes. Walmart, in fact, has examples of exactly this kind of improvement: store-manager shift planning moving from roughly 90 minutes to roughly 30 minutes per cycle. That is not nothing. It is genuine value capture, and it is worth doing. It is also, in a phrase that should set off mental alarms, retrofit.

A retrofit makes the existing operating logic run more efficiently. The work that used to take five people an afternoon now takes two people a morning, with help from agents. The substrate is preserved. The benefit is real. The benefit is also, depending on the workflow, captured by your competitors easily within twelve months (call that a rule of thumb, not a hard measurement), because the same retrofit they could do, they will do.

Reimagining starts from a different question. Instead of "how do we make this process faster," you ask: "what would this enterprise look like if I designed it today, with agents as a first-class participant in the workforce?" Not human work automated by agents. Work itself redesigned around what a hybrid human-agent organization can actually do.

The classic example is Block, the Jack Dorsey company that publicly announced the collapse of several layers of its management hierarchy in early 2026. The reason it could do that was not that agents made each individual manager more efficient. The reason was that the supervisory logic that required so many layers of managers in the first place no longer applied when agents could carry context, coordinate work, and make many of the small judgment calls middle managers had been needed to make. The reorganization wasn't a productivity move. It was a recognition that the substrate had changed.

A reimagine is also harder to copy than a retrofit. A reorganization, a redrawn workflow, a different set of approval layers - these take longer for a competitor to reproduce, because they require not just adopting a tool but rewiring an organization. The retrofit is twelve months to catch up. The reimagine can be three to five years, sometimes longer, sometimes never.

JPMorgan's onboarding redesign is a different piece of organizational machinery. The number of human eyes on a file, the sequence of steps, the way exceptions get handled, all of it gets redrawn. That is what reimagining looks like in operating terms.

I want to stress what reimagining is not, because I have watched executives nod sagely at this distinction and then proceed to retrofit aggressively while telling themselves they are reimagining.

Reimagining is not "we did our retrofit at a bigger scope." It is not "we adopted AI across many functions instead of just one." It is not "we are spending a lot on AI." It is not "we have hired a Chief AI Officer." None of those are reimagining. All of them are retrofits, possibly large ones. You can easily spend a billion dollars on AI without reimagining anything. You can also reimagine quietly, with relatively modest spend, by changing the way authority flows in a single critical workflow.

The diagnostic question is simple. After your AI deployment, is your enterprise running the same operating logic faster, or a different operating logic at a different shape? If the org chart, the handoffs, the layers of review, and the way decisions move through the company look fundamentally the same as before, you are retrofitting. If they look fundamentally different, you are reimagining. Most enterprises will do some of both, in different parts of the firm, at different times. That is normal. The point isn't to be one or the other. The point is to know which one you are doing in any given function, and to have made that choice on purpose.

Three caveats so this isn't misinterpreted.

First, this is not a binary. Every enterprise will retrofit some things and reimagine others, simultaneously. Customer support might get reimagined while finance gets retrofitted, or vice versa. Different functions sit at different points on the spectrum. That is normal.

Second, this is not a prescription. There are functions and industries where retrofitting is the correct answer, full stop. The argument is not that you should reimagine. The argument is that the choice should be made deliberately.

Third, reimagining is not safe. Neither is retrofitting. That is the next section, and it is the one most readers are going to be tempted to skim, which is precisely why I recommend you slow down for it.

There is no safe answer

If you finished the previous section thinking, "OK, retrofit is the conservative move, reimagine is the bold move, and conservative makes sense for an enterprise like mine," I have to interrupt your train of thought.

Both ends of the spectrum carry real and serious dangers. They are different dangers, and one of them is more visible in the public record than the other, which is part of why most executives have miscalibrated their sense of which is worse.

Retrofit's danger is strategic irrelevance. You preserve the working machine. You get more efficient. You keep the substrate intact. And in doing so, over a few years, you become a company whose competitive position is being eroded by competitors who didn't.

This is a slow death and likely a quiet one. It does not show up as a single bad quarter. It shows up as gradual margin pressure. As a steady drift in your win rates against newer competitors. As your customer base aging. As your best people leaving for firms that feel more like the future. As your retrofitting efforts producing real, measurable savings, which then get competed away on price within twelve months because everyone else can do the same retrofit.

The hard data is consistent with this picture. Gartner's forecast that more than 40 percent of agentic AI projects would be canceled by the end of 2027 is, on its face, a story about projects failing. It is also, more interestingly, a story about retrofits failing. The canceled projects are overwhelmingly retrofits that didn't pencil out. MIT NANDA's finding that 95 percent of enterprise AI pilots produced no measurable P&L impact tells the same story from a different angle. The pilots that didn't produce results were almost all retrofits.

To be clear about the provenance: neither firm labels the failures that way. The reading is mine. A pilot, almost by definition, slots the technology into a process the enterprise already runs, and genuine redesigns are rare enough that the ones we know about travel by name; you will meet most of them in the chapters ahead.

If you are an executive at a medium or large enterprise, you very likely have personal experience with quietly-failed retrofits in your own organization. You probably do not need to be persuaded that retrofit failure exists. You have lived it.

What you may not appreciate is that retrofit failure, at scale, is the more likely path to strategic irrelevance than reimagine failure is to immediate disaster. The slow bleed gets you. It gets you on the timeline of years rather than quarters, but it gets you. Quietly. Without anyone naming a moment of failure.

Now the other side.

Reimagine's danger is executional collapse. You tear up the substrate. You bet on a different operating logic. And the replacement cannot yet carry the load.

The canonical case here is Klarna, the Swedish buy-now-pay-later company, and its public reversal in May 2025. The arc is by now reasonably well-known but still sharp enough to be worth recounting briefly.

In February 2024, Klarna announced that an AI assistant was handling the work of roughly 700 customer service agents across 35-plus languages. Through 2024 and into 2025, the company ran a hiring freeze, with headcount falling from roughly 4,500 to roughly 3,500. The company's CEO, Sebastian Siemiatkowski, became something of a public face for the "AI as workforce replacement" narrative.

In May 2025, Siemiatkowski reversed publicly. In a Bloomberg interview on May 8, 2025, he said, with unusual candor:

"From a brand perspective, a company perspective, I just think it's so critical that you are clear to your customer that there will always be a human if you want."

And in the same interview:

"Really, investing in the quality of human support is the way of the future for us."

Klarna began rehiring and reassigning internal employees back to customer service work.

Then September 2025 happened. Klarna IPO'd, partly on the strength of the AI-efficiency narrative the CEO had spent the previous eighteen months building. The company opened its first day of trading at roughly $19.65 billion in valuation, 30 percent above its IPO price, and closed the day up about 15 percent. The AI walk-back had been public since May. The IPO investors knew about it, in some sense. They probably did not feel the full weight of it until subsequent reporting in September 2025 confirmed both the rehiring and the broader recalibration.

Investors learned that the company they had just bought had quietly walked back a significant part of its strategic AI thesis within roughly ten weeks of the offering. That is the consequence side of reimagine failure, made unusually visible by the IPO timing. Real money. Real accountability. Real loss of credibility.

Here is the dangerous thing about Klarna as a teaching moment. The data tell one story. Executives' fears hear a different one.

The data, as we have just walked through, all point the same way. The Gartner number is about retrofits. The MIT NANDA number is about retrofits. The vast majority of failed AI initiatives - the cancellations, the no-P&L pilots, the quietly-shelved projects in your own organization - are retrofits that didn't pencil out. Retrofit failure is everywhere. By the count, it is the dominant failure mode.

But retrofit failures don't have a face. They don't have a name. There is no widely-discussed magazine cover story about a Fortune 500 firm that retrofitted aggressively, captured eighteen months of efficiencies, and then watched its margins erode. The failures are aggregated into Gartner percentages and McKinsey buckets. They don't star in anyone's mental movie.

Klarna stars in everyone's mental movie. A Swedish unicorn. A named CEO. A confident pre-IPO narrative. A public reversal. A specific quote about humans and brand. An IPO that priced anyway. Subsequent reporting that confirmed the rehiring. The story has a beginning, a middle, and an end. The story has a face.

The human brain weights the named story over the unnamed aggregate. This is why Klarna shows up unprompted in many senior-executive conversations I have about AI failure, and the Gartner number shows up almost never. It is why "we should be careful about reimagining; remember Klarna" is something I have heard a dozen times in the past year, and "we should be careful about retrofitting; remember the 95 percent" is something I have heard close to zero times. The asymmetry is not in the data. It is in the storytelling.

And it bends executive intuition in a particular direction. Klarna makes reimagining feel risky and retrofitting feel safe. The data say the opposite. The data say retrofitting is where most of the failure lives. Klarna is one named, well-publicized event. The 95 percent is a thousand quiet ones, including, very likely, a few in your own organization that you have not yet labeled.

The evidence for retrofit failure is much denser than the evidence for reimagine failure. There is a Gartner number for retrofit failure. There is an MIT NANDA number for retrofit failure. There is at most a handful of named cases of reimagine failure, of which Klarna is the starkest (so far). The public record probably undercounts the rest, because large enterprises are reluctant to publicize internal reversals. The asymmetry is real, and it should make you suspicious.

Suspicious of what? Of the intuition that "we will be conservative about this" is a safe move. The data say it is not.

Both sides can hurt you. They will probably affect different companies in different industries on different timelines. The right position for your enterprise is not "conservative" or "aggressive." It is the specific position calibrated to your strategy and the outcomes you intend to deliver. The job of the rest of this book is to help you find and execute against that position.

There is one more thing I want to flag before this chapter ends. The named reimagine failures we have, Klarna foremost among them, are all visible because they happened. There is a class of failure we don't yet have on the public record: the enterprise that retrofitted aggressively, did everything textbook, captured real efficiencies, and then got outcompeted by a peer that reimagined. We don't have that case yet because the lag between disruption and visible documentation is real. The competitor pulls ahead, the gap widens, and three or four years later someone writes the Harvard case study.

My expectation is that we will start seeing those case studies by 2027. They will not be subtle.

Which leaves the question I cannot answer for you in this chapter: is your enterprise the case study someone else will write?

• • •

You now have the basic framework. You can see the retro-fit-to-reimagine spectrum. You understand the failure modes at both ends. You are starting to think about where on the spectrum your organization should sit, given your strategy and your appetite for risk. Maybe you are even feeling a little settled. The choice is hard, but at least it is a choice you can see clearly now, and you can start working it out with your team.

• • •

Key takeaways

- **Every senior executive is making a retrofit-or-reimagine choice right now.** Knowingly or not.
- **Retrofit failure is more common in the data.** Reimagine failure is the more memorable one.
- **A retrofit gets matched in roughly twelve months.** A reimagine takes years to match, sometimes never.
- **The discipline this book offers is choosing your position deliberately.** Not by drift, not by default.

• • •

Don't get comfortable.

Because the position you are about to pick is not fixed. There is a current running through this whole landscape, and it does not care where you decide to start. Think of it as a flowing river. Sometimes a raging river. It never stops.

Whatever location you pick today will be wrong tomorrow.

The question is not where to stand. The question is how to move.

Further reading

Companion website: riverdoesntwait.com

A few readings to go deeper on this chapter. Sources covered in earlier chapters are not relisted here.

Sebastian Siemiatkowski / Klarna, Bloomberg interview, May 2025

The reversal interview is the centerpiece, but the broader arc is what the chapter relies on: the February 2024 announcement that AI was handling the work of roughly 700 customer service agents, the May 2025 Bloomberg interview where Siemiatkowski reversed publicly, the September 2025 IPO that priced anyway, and the subsequent reporting that confirmed the rehiring. Reading the sequence in order rather than the reversal moment alone gives you the felt experience of how reimagine failure unfolds when it unfolds in public.

Chapter Three

Reading the River

Why navigating agentic AI is a continuous challenge, not a one-time decision

A river is a useful image to end a chapter on. It is a less useful image to start one with, because by the second sentence the metaphor risks getting cute. So I will be careful with it. But I want to keep it because the thing the river is doing is something the rest of the book has to navigate around.

A river has three properties that matter here.

- First, it never stops moving.
- Second, it almost never reverses direction.
- Third, its pace varies. Sometimes it flows gently. Sometimes it surges hard enough to capsize what was peacefully navigating it the day before.

The current of this "river" of agentic capability has all three properties. That is the argument of this chapter. It is also the reason the position you settled on at the end of Chapter 2, somewhere on the spectrum between retrofit and reimagine, will not stay correct. The position will move because the river will move it. Your job is to learn how to effectively navigate this river.

The current

Foundation AI models keep getting more capable. Agents keep getting more useful. The benchmarks keep moving. None of this is news to you. What is worth pausing on is the structural effect on the master choice between retrofit and reimagine that we just spent a chapter establishing.

As capability advances, two things happen at the same time. One is that the pressure to reimagine intensifies. Competitors who reimagine pull further ahead on unit economics, on cycle time, on operating leverage. The retrofit you ran last year still works, but the gap between you and the firm that redesigned the workflow widens. The danger of strategic irrelevance, which we just established was real, gets worse as the technology gets better.

The other thing that happens is that the capacity to reimagine improves. Agents that could not handle ambiguous, multi-step, judgment-laden work last year can handle more of it this year. The work you wanted to redesign last year but could not, because the agents could not yet carry the load, becomes redesignable. The danger of executional collapse lessens as the technology matures.

But the risk of total disruption rises as competitors are also enabled by the same improving technologies.

Both effects pull in the same direction. Both effects compound. The pressure to reimagine grows. The capacity to reimagine grows alongside it. The right position on the retrofit-to-reimagine spectrum, whatever it was for you twelve months ago, has already moved further toward reimagine than it used to be. Twelve months from now it will have moved further still. And probably at an accelerating pace.

That is the current in this river of agentic AI. It is not just metaphorical. It is the structural consequence of capability advancement on a master choice that exists on the retrofit-to-reimagine spectrum.

A current does not move at a constant speed everywhere. The same river runs slowly through a wide flat stretch and fast through a narrow gorge. The same is true of the current of agentic capability. Some functions and some industries feel the pull harder than others.

The current moves faster where the enterprise problem is knowledge fragmentation, repetitive exception handling, or coordination across many systems. Telecom customer care. Bank operations. Proposal generation. Logistics validation. Document drafting. Employee knowledge search. In those waters, the current is strong, and the firms that have not yet adapted are visibly drifting backward against newer ones that have.

The current moves slower where error costs, human trust, or empathy dominate. Clinical encounters. High-stakes advice.

Lightly structured human interactions where the texture of the conversation is the work. In those waters, the current is real but the flow is gentler, and the cost of moving prematurely is higher than the cost of waiting another year.

That variance is not counter-evidence to the current. It is evidence that river speed varies by channel. The river is moving in all of these places. It is just moving at different rates. And it is, on average, accelerating.

When the river surges

A river also does not move at constant speed over time. Sometimes it flows gently for months. Sometimes a snowmelt or a storm changes the math overnight. The water that was content to glide through a culvert yesterday is, this morning, taking the culvert with it.

The current of agentic capability has the same kind of behavior. Most of the time, it is incremental. Benchmarks move a few points. A model release happens that is a little better than the one before. A platform vendor adds a feature. The improvements are real but you do not feel them as discontinuities. The river is steady.

And then sometimes it surges.

In the second week of April 2026, within roughly a fortnight, two frontier labs released AI models that visibly raised the ceiling. Both shipped at one-million-token context windows (which basically means that these systems were able to keep in their short-term memory about a million words' worth of text - equivalent to about

20 copies of the book you are holding). Both were positioned, in their public framing, not as conversational assistants but as agentic platforms. Infrastructure for autonomous work. On OSWorld, a benchmark designed to measure agentic computer use on tasks an office worker would recognize, scores roughly doubled in the eleven months from May 2025 to April 2026.

That is what a surge looks like. It does not announce itself with a press release labeled "surge." It announces itself with two model launches in two weeks and a benchmark line that bends hard upwards.

A historical example, for pattern. In September 2025, Anthropic released Sonnet 4.5 and demonstrated it autonomously rebuilding a working version of the Claude.ai web application over roughly five and a half hours, using more than three thousand tool calls. That demonstration was a surge moment. Long-running agentic workflow stopped being aspirational and started being a thing you could watch on a recorded screen. The benchmark of "what an agent can do without supervision for an extended period" moved in a single afternoon. Not because the model alone was better, but because the model met a harness (a way of controlling the model), and the harness met a public demonstration, and the public demonstration moved the conversation.

That is the underrated part. Surges are not just model-release moments. They are moments when reasoning improvements meet distribution, orchestration, and governance surfaces. The model alone, sitting on a server somewhere, is necessary but not sufficient. The surge happens when the model becomes a thing your organization can deploy, against work it can structure, with controls it can defend. When all three conditions arrive in the same week, the river can jump its banks.

Some surges are visible only when they land in the real world. The model gets released, the benchmark moves, and you read about it on a Tuesday morning before your operating committee meeting.

Other surges are visible before they land. There is a pattern that has emerged at the frontier labs of deliberately holding back capabilities the labs themselves consider too dangerous to release as-is. A model trained but not deployed. A capability demonstrated internally but withheld from easy integration into other systems. The fact that the labs are doing this, talking about doing this, and publishing about doing this, is itself a surge signal. Capability that exists privately will eventually be released publicly, in some form, on some safety timeline. When you read that a frontier lab has held back a model, you are reading the announcement of a surge that has not yet landed but is on the way.

Take Anthropic's Mythos, the plainest example of this pattern as of June 2026. Anthropic trained the model. Anthropic chose not to release it. The stated reason, in essence, was that the model's capability profile, particularly on a class of tasks the lab judged could create disproportionate downside risk if released into systems that hadn't yet learned to defend against them, had outrun the public defenses around those applications.

Read the structure carefully. The model existed. The harness to deploy it was within reach. What was missing was the third condition, the governance surface, the public defenses around the most sensitive applications. Anthropic looked at the three conditions and saw them about to converge, and chose to wait for the third (governance) to mature. They saw the river about to jump its banks and decided to wait for the banks to be reinforced first, or at least for those navigating the river to prepare for what was coming.

Just as this book reached final editing, Mythos was released as Fable 5, with significant restrictions on how the model can be used in the areas of cybersecurity and biological research. The pattern, lab foresight as a surge signal, is what matters, and the pattern will keep producing examples.

A surge is the dramatic case, the one that can capsize you, which is why it gets the attention here. But the current's pace moves both ways. Every so often something upstream pushes back and slows it: a regulator that wants new models reviewed before they ship, a limit on compute or data, a safety scare that makes the labs more cautious for a while. The water does not reverse, because capability that already exists does not vanish, but for a stretch the current can run slower than the trend would predict. A slowdown like that is a lull, and almost always a temporary one; the current picks back up when the obstruction clears. The discipline it calls for is the opposite of relief. You use the easier water to let your slower outcomes catch up, and you keep watching, because the lull does not last, and the river does not wait for anyone who mistook it for a stop.

What matters for now is that the river's pace varies, that surges and the occasional slowdown are both part of how the current behaves, and that watching for them is part of the executive's own navigation work. The watching is not open-ended 'AI-spotting.' It is bounded to the work behind the outcomes you actually own. We will come back to how that watching looks later in the book.

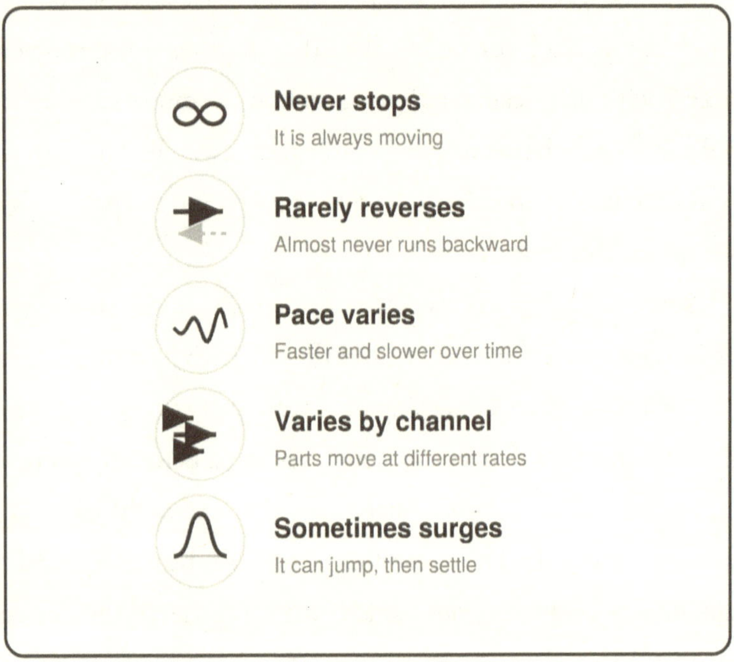

Characteristics of the 'river' of agentic AI

• • •

The river is real, but the boats are not all floundering

The version of this argument every vendor pitch tries to ignore is the one that matters.

The current is real. But it is not all-conquering. Not yet, anyway, and possibly not ever.

You read in Chapter 1 about the METR study showing experienced developers were slower with AI tools than without them. Hold that finding in your head while I make this point. Capability advancement is genuine. The integration cost of capturing that capability inside a working enterprise is also genuine. Skilled humans on familiar work, working in tools they already know, do not automatically benefit from a more capable agent showing up next to them. The gap between "the technology can do this" and "our organization can do this with the technology" is real, persistent, and frequently larger than the buyers of the technology expect.

That is true, and yet the river is still flowing and rising. The current is the structural force. The friction inside your enterprise is the local resistance. Rocks or even boulders in the river that disturb the flow and sometimes cause rapids. The current does not care about local resistance, in the long run, but in the short run the local resistance is most of what determines whether you capture value from this year's capability advance or watch it pass through you on the way to a competitor.

Or, to put it another way: the primary bottleneck halting the deployment of agentic AI in production is not a lack of access to frontier models. It is the crushing weight of legacy technical debt, inadequate enterprise data governance, and a systemic inability to integrate modern interfaces into archaic, decades-old processes. Renting access to the AI model is the easy part. Everything around the model is the hard part.

The current keeps moving. Your organization's ability to ride it is constrained by things that have nothing to do with the model and everything to do with the substrate the model has to run on top of.

That tension is most of what the rest of this book is about.

Navigation, not selection

In Chapter 2, I asked you to think about where on the retrofit-to-reimagine spectrum your enterprise should sit. That was a useful framing, because the spectrum is real and the choice you are making, deliberately or by default, is real. But the framing was, by design, partial. It treated the spectrum as a static thing on which a position could be selected. That is not how the spectrum actually behaves.

The right position on the spectrum is not selected. It is navigated.

There is a difference. Selection assumes a stable choice. Navigation assumes the answer keeps moving. Selection rewards being right. Navigation rewards being early to notice that you were no longer right. Selection looks like a strategic-planning exercise, conducted once, defended afterward. Navigation looks like an operational discipline, conducted continuously, with a mechanism for noticing when the assumptions underneath the position have shifted.

The reason the executives running ahead are running ahead is not that they picked a better spot than you did. It is that they figured out, faster than you did, that the spot was going to move, and they built the capability to move with it. The retrofit they ran last year worked because last year was a retrofit year. This year they have the conviction to redesign the workflow that retrofit was built on top of, because this year is a redesign year. Next year they will be doing whatever next year asks for, because they have the operating discipline to ask the question continuously rather than pick an answer once.

You may have noticed that the executives we keep quoting in this book talk about agentic AI in the present, continuous tense. They are not announcing a destination. They are describing a trajectory. Dimon's 2024 letter compared the moment to the printing press and electricity, framed as augmentation. His 2026 letter says AI will affect every function, application, and process, framed as redesign. The same firm. Two years apart. The position changed. The institution moved with it.

Dimon also said this, in early 2026: "The pace of adoption will likely be far faster than prior technological transformations, like electricity or the internet. Those took decades to roll out, but this implementation looks likely to accelerate over the next few years."

Read that quote not as a prediction. Read it as a description of the current. He is not telling you what is going to happen. He is telling you what he is already navigating.

This is the move I am advising you to make. Stop asking "what is the right position for my enterprise on the retrofit-to-reimagine spectrum." Start asking "what mechanism does my enterprise have for noticing when the right position has changed, and for moving when it has." The first question has an answer that goes stale. The second question has an answer that gets sharper with use.

The rest of this book is about the second question. Specifically, the rest of this book is about the three operational levels (I call them 'Tensions') at which the navigation actually happens.

Inside the boat

Three Tensions. Workflow and workforce. Context. Supervision, autonomy, and governance. Each one is a chapter. Each one is a place where the master choice we just established gets made operationally, in real decisions, by real executives like you, with real consequences. Each is a spectrum with a pole at either end, and the work is not to pick a pole but to choose where along it a given piece of work belongs. That position is its setting, and navigating is the work of moving the setting as the river moves.

Up to this point the chapter has been about the river. The river is the external force you are navigating. Continuous, accelerating, indifferent to where you sit. The three Tensions ahead are something different. They are not more river. They are what you have to get right inside the boat to navigate the river competently.

The boat is whatever is within your domain of control. It might be the entire enterprise. More often, it is a function, a division, a business unit, or whatever scope of responsibility you actually hold. Your boat is that thing. The three Tensions ahead are decisions about that thing.

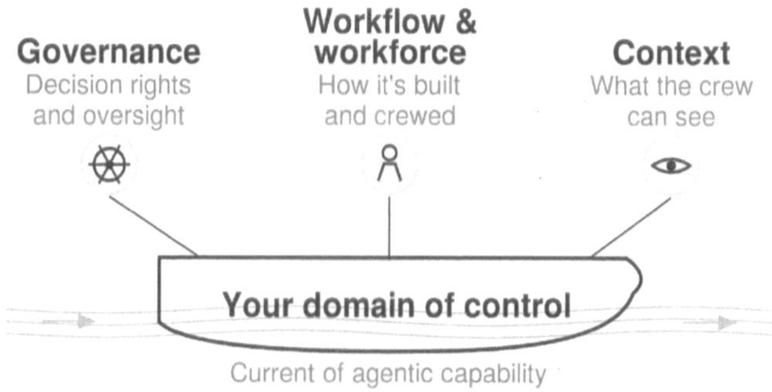

Navigating your organizational 'boat' via setting the three Tensions

What sits beneath the three Tensions is the work you covered in the previous chapters. Not a catalog of what agents can do, which you will build up incrementally as you read these chapters. Something more basic. The recognition that the master choice (retrofit or reimagine) is real, that you are making it continuously whether you realize it or not, that neither pole is safe, and that the right answer is navigated rather than picked. Without that posture, you never engage the first Tension as a real choice. You just put agents into existing seats and call it AI adoption.

That is what was happening in many of the failed pilots we discussed in Chapter 2. They retrofitted because they treated their part of the organization as a fixed thing and the agent as a new occupant of an existing seat. They never asked whether the boat itself should be designed and run differently.

The first Tension is workflow and workforce. The design of the boat and who staffs it. Once you accept that the boat itself can be reshaped, the question becomes what work you let agents do, and how that work is organized. Do you slot agents into existing roles in existing workflows, with the same approval layers and the same handoffs and the same supervisory hierarchy? Or do you redesign the workflow and the roles inside it together, on the assumption that a hybrid human-agent organization can do work the all-human version of the same organization could not? The pole on one side is "agents take seats." The pole on the other side is "the seats themselves are different." This is the most concrete and immediate of the three Tensions, which is why we tackle it first.

The second Tension is context. As the boat is being redesigned, the question becomes what each part of its crew (whether human or agentic) can see. Agents likely need richer information to do the new work, which is most of what context provision is. How much organizational context do you give them? How broadly does that context extend? Inside one function, across the enterprise, or across organizational boundaries to partners, customers, suppliers?

The pole on one side is "tightly bounded context, narrow agent." The pole on the other is "broad context, agent operating across the seams of the organization." Context is harder than it sounds. It is also, I think, the most interesting Tension in this book, because the scope of context you own turns out to be roughly the same thing as the scope of organizational influence you hold. Making that context explicit and handing it to a team means giving up your exclusive hold on the knowledge your influence is built on. That makes context provision politically and psychologically uncomfortable in ways that have nothing to do with the technology.

The third Tension is the cluster of supervision, autonomy, and governance. The decision rights aboard the boat, and what is built into the substrate that catches a bad call before it sinks the whole thing. As the work is being redesigned and the context being provided, the question is how much of the resulting work the agent gets to do without a human in the loop, and what mechanism keeps the autonomy from going off the rails. Autonomy without governance is reckless. Governance without autonomy is theater. The two questions are joined at the hip, which is why the chapter combines them.

The pole on one side is "tight human supervision, narrow autonomy, governance applied as a brake." The pole on the other is "wide autonomy, governance built into the substrate so the agent's default behavior is the safe behavior." This is the hardest of the three Tensions to hold at the right altitude, because it pulls toward either generic AI ethics slideware or technical implementation, neither of which would help you make this choice.

The three are tightly linked

Workflow redesign is what creates the need for richer context. If the workflow is unchanged, the context the agent needs is whatever existed in the old workflow, which is a known quantity. The moment you redesign the workflow, you are operating with new information requirements, new handoffs, new dependencies. Context becomes an open question.

Context, especially context that includes the desired outcome rather than only the procedure, is what makes meaningful autonomy possible. An agent operating against a tightly specified procedure does not need much autonomy. An agent operating against a desired outcome with the latitude to find the procedure that gets there does. The richer the context, especially context about goals and constraints, the wider the autonomy that becomes safe to grant.

And autonomy is what raises the governance question. A tightly supervised agent operating against a narrow procedure does not need elaborate governance, because there is no room for the agent to do anything dangerous. The wider the autonomy, the larger the surface on which something can go wrong, and the more the question of whether governance is bolted on after the fact or built into the substrate becomes the difference between operating credibility and operating crisis.

You cannot navigate any of the three Tensions in isolation. Workflow without context goes nowhere. Context without workflow redesign is wasted on agents the workflow won't let do anything interesting. Autonomy without context is reckless. Governance without autonomy is theater. The three settings get made together, in unison, for any given piece of work.

This is also why the book does not give you the three Tensions in alphabetical order, or in order of importance, or in order of urgency. The executive working on a piece of real work in front of them does not set the three Tension settings one at a time. The operational reality is that the three settings get made together.

What comes after the three

There are two more chapters after the three Tensions, and I want to flag both now so the reader who skims ahead finds what they expect.

The first is a chapter about what to do now. It pulls together everything you have just read about retrofit and reimagine, the current and the river, and the three Tensions, into a discipline you can actually run with your team. It is also where the watching discipline I promised earlier lives, because that watching is operational work and it belongs alongside the rest of your team's operational work.

The book then closes with a chapter that is not a Tension and is not directional. A chapter about you. What does all of this mean for your role, your career, your sense of what executive work actually is? If the choices we are about to walk through change the shape of the enterprise, and you are an executive in an enterprise, they are also going to change the shape of you. For now, what I want you to know is that the book ends with something personal rather than something purely analytical. The choices the rest of the chapters describe ultimately point back at the person reading the chapters. At you. That is the intended shape.

• • •

Key takeaways

- **The river has a current of agentic capability that never stops and rarely reverses.** Its pace varies by channel and over time, mostly accelerating but now and then slowed for a stretch by forces outside the labs.

- **The right position on the retrofit-to-reimagine spectrum is not selected; it is navigated.** The river keeps moving, so the right position keeps moving with it.

- **Surges occur when improvements in AI reasoning meet distribution, orchestration, and governance surfaces within the same window.** Each surge is a convergence, not a simple release.

- **Three Tensions sit inside the boat.** Workflow and workforce. Context. Supervision, autonomy, and governance.

- **The three Tensions are linked.** Their settings move

together as the work moves further toward reimagine.

• • •

So the framework is clear. The current is real and the river is moving ever faster. And the three Tensions ahead describe the settings with which you control and navigate your boat.

Now we begin the work itself, starting with the most concrete and immediate of the three. The agents are ready to come aboard. The first question is whether you put them in the seats you already have to do the work you are already doing, or you redesign the seats and the work together.

Further reading

Companion website: riverdoesntwait.com

A few readings to go deeper on this chapter. Sources covered in earlier chapters are not relisted here.

Anthropic's Sonnet 4.5 release and demonstration, September 2025

The historical surge anchor in this chapter. The released demonstration of the model autonomously rebuilding a working web application over hours, with thousands of tool calls, is the best single artifact for what "long-running agentic workflow" looks like in practice (at the time of writing). Worth watching the recording rather than reading about it; the duration is part of the point.

Anthropic's public statements on Mythos

The held-back-frontier-model anchor in this chapter. Anthropic's communication about why it initially chose not to release Mythos is the best current example of lab foresight functioning as a surge signal. The situation resolved just as this book reached final editing, with the model released as Fable 5 under significant restrictions; the lab's public framing of the decision remains the artifact worth tracking, not the specific resolution. Watch how the lab described what condition was missing and what was put in place before release, because that condition is the one your enterprise will eventually have to navigate around.

OSWorld benchmark

The benchmark that produced the rough doubling this chapter cites in its discussion of the April 2026 surge. OSWorld is one of the more honest agentic-capability measurements in the public research base, in that it treats the agent as a worker operating against representative computer-use tasks and reports failure modes alongside successes. Useful counterweight to vendor benchmark cards. The leaderboard updates over time, so the headline numbers in this book will lag the current state by the time you reach this section.

UK AI Security Institute, Frontier AI Trends Report

A government-led perspective on where frontier capability is going, with special attention to the kinds of dangerous-capability thresholds that drive lab decisions to hold models back. Reading this in parallel to lab-published material gives a sharper picture of the surge-signal pattern than either source alone. The argument in this chapter that lab foresight is itself a surge signal lands more concretely once you have read what governments are watching for.

Chapter Four

Tension 1: When the Work Itself Changes

Why slotting agents into existing roles is the slow path to irrelevance

There is one operational decision in agentic AI that matters more than any other. Most senior executives believe they are making it. Most aren't. They are making something that looks like it from a distance: deciding which agents to deploy and where, evaluating vendors, approving pilots. They are missing the underlying choice the deployment is forcing on them.

The choice is whether workflow and workforce are one decision or two.

Most enterprises are treating them as two. The IT side scopes the agent. HR sizes the team. The COO redesigns the process.

Each function makes its own move on its own schedule. By the time someone looks at the resulting whole, the workflow has been sped up in places, the workforce has been resized in others, and the relationship between the two looks accidental.

The enterprises pulling ahead are treating them as one. They redesign the workflow and the roles inside it at the same time. They change what the work is and who or what does it together. The two questions are really a single question wearing two outfits, and the answer is what your boat's design actually is.

Microsoft's 2026 Work Trend Index surveyed more than nineteen thousand workers to find out which factors actually drive AI impact at the enterprise level. The result: organizational factors (culture, manager support, talent practices) matter about twice as much as individual factors like mindset and behavior. The constraint sits in the system. Which means that fitting more capable people into an unchanged system, or better agents into an unchanged workflow, will not produce the results executives keep expecting.

This is the first of the three Tensions, and it is the most concrete and operational of them. It is also the one whose financial effects are the most directly measurable right now: workflow redesign shows up immediately in cycle times, headcount, and cost, while the payoffs from the other two Tensions are just as real but can be slower to put a number on. The choice between treating agents as new occupants of existing seats and treating the seats themselves as up for redesign is the substance of this chapter.

One decision, not two

Workflow and workforce are one decision, not two. Saying it is easy. Acting on it is harder, and the reason has more to do with org structure than with conviction. In any large company, workflow lives with operations, workforce lives with HR, and the technology that touches both lives with IT. Three different functions, three different review cycles, three different ways of measuring success. Two of them, IT and HR, often don't even sit within your direct scope; they are centralized functions you have to negotiate with. Even when you can see the pieces, you may not control them all. Each can make a move that looks correct in isolation. The combined result can still be incoherent.

The entanglement runs in both directions, and that matters for diagnosis.

Going from workflow to workforce: the moment you redesign how work moves, you have changed who needs to be in the workflow and what they need to be doing. New handoffs require new roles. Compressed cycle times require different escalation paths. Say an agent now does the routine 90 percent of a process. The human role that survives concentrates on the harder 10 percent: the exceptions, the judgment calls, the cases the agent can't handle. That is a different job from the one that owned the whole process before, and the workforce decisions have to follow. The workforce question is downstream of the workflow question, but it is not optional and not secondary.

Going from workforce to workflow: the moment you change who or what is doing the work, you have changed what the workflow can plausibly be. An agent placed inside a process designed around a human will perform like a faster, cheaper, potentially slightly less reliable human. An agent placed inside a process designed around what an agent can actually do can compress steps, change sequence, and eliminate handoffs that existed only because a human had to wait for another human. The same agent, dropped into two different workflows, produces two different outcomes. If your enterprise has been getting underwhelming results from agents, look at the workflow they were dropped into before you blame the agent.

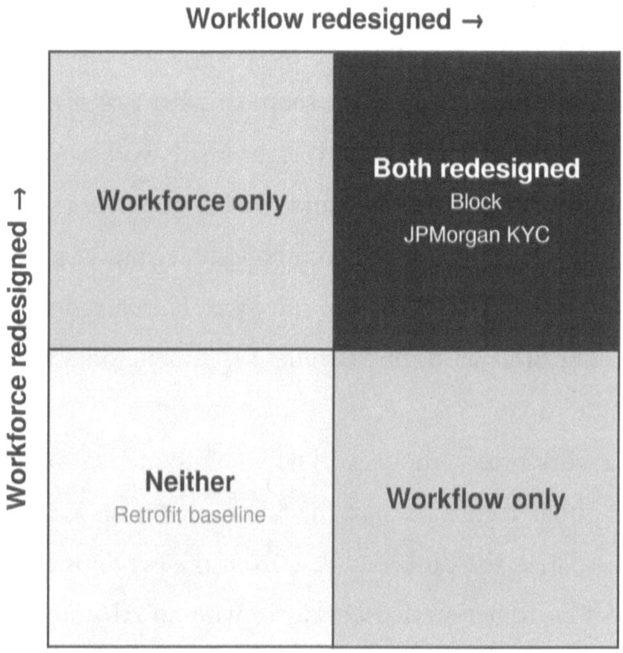

Workflow vs. workforce redesign. True reimagining requires both

This discipline is older than the technology now forcing it. In 1990, Michael Hammer wrote one of the most-cited articles in business history under the title "Reengineering Work: Don't Automate, Obliterate." His central argument was that putting computers on top of existing processes was a category error (a mistake about what kind of problem you were solving). The processes themselves had been designed around the limits of the previous technology, and automation that preserved those limits captured a fraction of what was actually available. Ford's accounts-payable example, the historical anchor of the article, eliminated about three-quarters of the function's headcount by redesigning the workflow such that invoices were no longer needed at all. The technology was unremarkable. The redesign was the move.

What is different now is the scope of what is available. Agents can carry context (next chapter), coordinate work across systems, escalate when they hit edges, and operate at machine speed continuously. Hammer's argument applies to agents at a higher level than it applied to enterprise software. The category error gets larger. The firms acting on Hammer's argument today are the ones pulling ahead.

The workforce-without-workflow failure mode is more visible in the public record because the failure shows up as a public reversal. Klarna, the canonical case from the previous chapter, replaced a customer-service workforce with an AI system without the workflow redesign that would have made the smaller workforce viable, and the company ran into trouble it has since acknowledged publicly.

This is the anatomy of what Chapter 2 called executional collapse. Klarna did not fail because it chose the reimagine end of the spectrum. It failed because it ran half a reimagine. The workforce changed. The workflow did not. And the replacement, dropped into a process still designed around humans, could not carry the load. That reading should sharpen, not soften, the Chapter 2 point: the case that makes reimagining feel dangerous is, on inspection, a case about doing it halfway.

The contrast cases are the enterprises that did the workflow redesign alongside the workforce change, kept the affected workers inside the organization in different roles, and have so far not produced public reversals. The next section walks through what that looks like in practice. The workflow-without-workforce failure mode is harder to see because it is quieter, and I will come back to it in the retrofit section.

When the boat itself changes

When enterprises actually reimagine the work rather than retrofit it, the redesign takes one of two distinct shapes. They produce different compression patterns - one thins the management structure, the other shortens the process itself - and demand different leadership moves, and smashing them together into one "reimagine looks like X" framing is how executives end up running the wrong play for their context.

Decision-structure redesign collapses the org chart itself. Layers compress. Decisions move to where the information actually is, rather than waiting for the information to climb a ladder. Operational-process redesign leaves the chart mostly intact and rewires the work happening inside it. Different handoffs, different decision points, different sequence. Both are reimagine. Both involve workflow-and-workforce coupling done deliberately. They just sit at different levels of the firm.

Decision-structure redesign

The cleanest public case is Block, the Jack Dorsey company. Dorsey has been on the public record since early 2026 about what he is doing, and the framing has been consistent. In his telling, there are currently roughly five layers of management between him and any other employee at Block, which employs about six thousand people. His target inside the next year is two or three. His stated ideal, in the most ambitious version of the framing, is none. "In the most ideal case," he said on a podcast in April 2026, "there is no layer."

The operational mechanism that makes that plausible at scale is what he calls the "intelligence layer." The majority of work at Block, he claims, now moves through agentic systems rather than between humans up and down a hierarchy. The agents do the coordination, context-carrying, and routine judgment work that middle management used to do. Humans, in the framing, concentrate at three roles: individual contributors who do the work, directly responsible individuals who own outcomes, and what Dorsey calls player-coaches who do both at once. The hierarchy collapses not because people are being told to do more, but because the work that the hierarchy existed to coordinate is being coordinated by something else.

The technology is doing less of the work than the framing implies. Block has publicly acknowledged that about 95 percent of code generated by their AI systems still requires human modification. The reorganization is real and the framing is bold, and the technology is also more partial than the headline numbers suggest. Both can be true at the same time.

I am not arguing that Block has built the future. I am arguing that Block has run a decision-structure-redesign play, has been transparent about what the play is, and is therefore the most compelling public case (as of this writing) of what reimagine looks like when it reaches all the way to how the company makes decisions, not just how it does the work. Worth pausing on what Block actually reimagined. It is not the path to running a management hierarchy more efficiently. The company changed what it was organizing around, from a hierarchy to what it describes as an 'intelligence.' That is a reimagined destination, not a reimagined route to a fixed destination.

Coinbase ran the same play three months later. In May 2026, the company announced a 14 percent workforce reduction along with the explicit elimination of what CEO Brian Armstrong called "pure managers" and a stated cap of five management layers below the C-suite. Armstrong's framing matched Dorsey's. Managers are expected to contribute individually as well as coordinate, with as many as fifteen direct reports each. Two named cases at the same pattern in the same quarter is enough to call it a pattern, not an idiosyncrasy of one founder.

What gets squeezed, in decision-structure redesign, is middle management. The layer whose job was carrying context across functions, applying judgment at the seams, and coordinating handoffs is exactly the layer that agentic systems are now best at replicating at scale. The work that defined the role can be substantially absorbed by software. What remains for humans in middle is concentrated, exception-heavy, and considerably smaller than the layer it replaces.

Operational-process redesign

The best-documented public case is JPMorgan's KYC system, the know-your-customer process that ingests, classifies, validates, and onboards new clients across the bank. The headline number, well covered in the trade press, is that the agentic system that went into production in April 2026 compressed a workflow that used to take five days into a workflow that takes under a minute. What the headline number leaves out is the workforce side of the same operation. The two hundred people who used to do the controls review for those onboardings were not eliminated. They were redirected to a different controls function inside the bank, doing work the agentic system cannot do on its own. The five-day-to-one-minute number and the two-hundred-person redesign are the same operation viewed from two angles.

The org chart at JPMorgan did not collapse. The bank still has the same overall shape. What changed is what happens inside the operational unit: different handoffs, different decision points, agents doing the routine validation work and humans concentrating at exception and oversight edges. The redesign is narrower than Block's. The org chart is roughly intact; the work inside one of its boxes has been completely rewired.

Mary Erdoes, who runs JPMorgan's Asset and Wealth Management (AWM) division, described a parallel pattern at the bank's Company Update in February 2026. A controls function that had used about two hundred people as the pilot is now being scaled across the firm: three thousand people inside AWM doing the redesigned work, with another three thousand to five thousand identified for the same treatment next. Erdoes called the eliminated portion of the old work "no-joy work," the kind of repetitive checking-and-routing labor that consumes time without producing satisfaction. The framing is corporate-positive and probably also accurate.

Jamie Dimon was direct about what this means at the same event. "We have displaced people from AI," he said. "And we offered them other jobs." The full framing in context: "We already have huge redeployment plans for our own people, in fact we spoke about it today, then we have to up that a little bit." The three pieces matter together. Displacement is happening. Redeployment infrastructure is in place at scale. And the bank is planning to expand the redeployment further as the redesign continues. This is workflow-and-workforce coupling done deliberately and at scale, by an institution with the operational discipline to do both parts of the move at the same time.

Cleveland Clinic ran the same shape of play on a different process. The health system redesigned its patient referral workflow, moving from a process that took multiple business days to one where most referrals route through the system in under a minute. The cycle-time figure is a vendor claim and worth flagging as such. The structural point is independent of the exact number: the workflow was redesigned, the roles inside it were redesigned, and the operational unit is in a pilot-moving-to-broader-rollout phase rather than at full health-system scale.

What gets squeezed, in operational-process redesign, is mid-tier execution. The work that sits between fully routine and fully judgment-driven. Junior tier-one execution often goes through agents entirely. Senior judgment work, the kind that determines what gets escalated and how exceptions get handled, stays human. The squeeze falls on the middle of the work itself, the layer where routine and judgment used to mix and where the routine portion can now be peeled off and given to software.

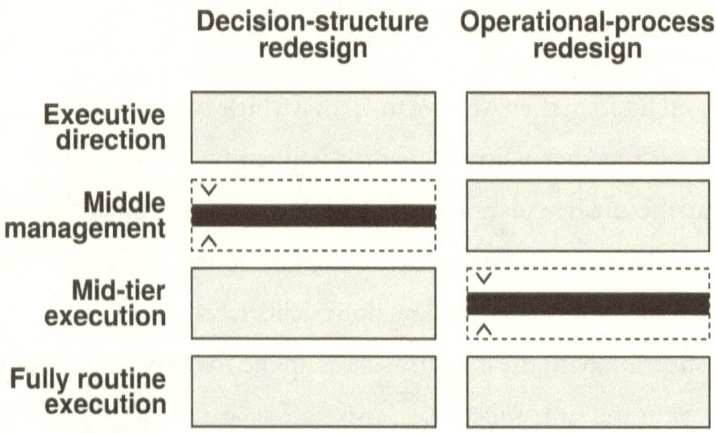

Different types of redesign result in compression at different organizational layers

The squeeze in the middle of the work

The two compression patterns matter for diagnosis. If you are running a decision-structure-redesign play, the layer that will start to look thin first is middle management. If you are running an operational-process-redesign play, the layer that will compress first is mid-tier execution. Same coupling principle, different surfaces.

The broader research on AI and work is convergent at the right level of analysis. The Federal Reserve Bank of Atlanta's March 2026 survey of corporate executives found that AI is reallocating tasks across firms rather than eliminating aggregate employment, with routine clerical activities exposed to substitution while analytical, technical, and managerial tasks more often described as being complemented. The paper does not use the specific "middle of the work" framing, but the empirical pattern is in the same direction.

The compression showing up in the company cases above appears across the whole labor market too: total employment holds roughly steady while a great deal of reshuffling happens underneath it. I am working at the operational level here, where the named enterprises sit; the economy-wide picture is a different conversation, held in different rooms by different people. The two pictures do not contradict each other.

The retrofit story

Retrofit is the simpler play and the one most enterprises are still running. It leaves the boat the same shape and adds horsepower to the engine. Workflow stays. Role structure stays. Approval layers stay. Agents arrive and start doing work humans used to do, sometimes faster, sometimes cheaper, sometimes both. The numbers come in. The savings are real. The implementation is comparatively quick because nothing organizational has to change.

Two things are worth being direct about, because most executives have been measuring one of them and seeing the other without quite registering it as part of the same story.

The first is cost-out. Headcount growth slows. Exception handling moves to a smaller team. Compute starts to do what people used to do. The savings flow to the operating margin. This is the part everyone has been measuring, and it is where the retrofit-or-reimagine ROI conversation has been concentrated. (Of course, a frequent failure mode is when the true costs of AI deployments are not accounted for, perhaps because they are held in a separate IT budget.)

The second is speed-up. The same workflow, run with agents inside it, often runs faster. Walmart's shift-planning workflow that compressed from about ninety minutes to about thirty per cycle is a clean example. Cycle times that used to take days now take hours. Exception handling that used to take hours now takes minutes. Customer onboarding that used to require five business days now requires one. Inside the operating unit, work simply moves faster.

Speed-up is genuine value and easy to underweight. An enterprise that captures a 40 percent speed gain on a high-volume workflow has produced something operationally meaningful even if the headcount line never moves. Speed-up is, in many cases, what retrofit actually delivers more reliably than cost-out. Worth measuring both.

What retrofit does not do is change the decision structure. Same decisions made by the same or similar people, same sequence. The work happens faster. The questions that get asked, the layers that approve them, and the order in which they are answered all stay the same. That is the structural limit of the retrofit play, and it is the reason retrofit advantages tend to compete away.

Retrofits that capture a 40 percent speed gain are not durable competitive moats. The same agents are available to your competitors. The same workflow patterns can be copied. The retrofit you ran this year, your competitor can run next year, and the gap closes. (This doesn't mean you shouldn't do them, it just means that they don't create sustainable advantage.) Reimagine is harder to copy because copying it requires rewiring an organization rather than adopting a tool. The duration of any specific reimagine advantage depends on the organization being copied, the path being blazed, and how much of the work is repeatable. Once a path is blazed, others can move faster than the original move took to execute. The structural advantage of reimagine over retrofit lives in this kind of difficulty.

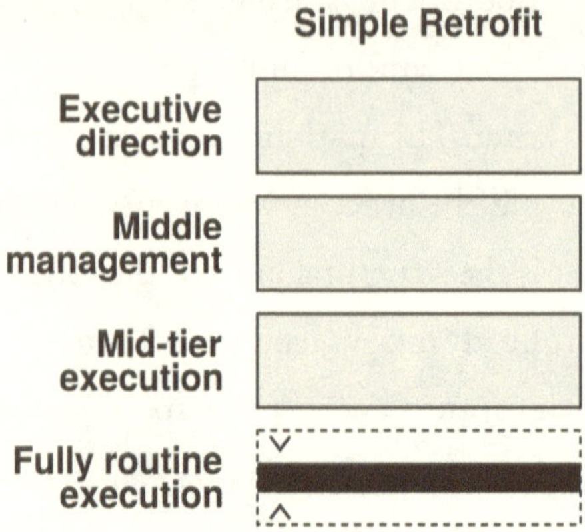

Simpler retrofit approaches generally compress only the bottom layer of organizations

The workflow-without-workforce failure mode shows up here, and we do not yet have a clean named example for it within the agentic-AI window. The pattern is structurally predictable. Workflow automation that runs without coupled workforce redesign produces real first-year gains and then erodes as the surrounding organization adapts back to its original shape. The technology delivers what was promised. The workforce structure around it stays designed for the workflow that the redesign was supposed to retire, and the people inside it quietly absorb the value the redesign was supposed to capture.

The reason we do not yet have a clear public example is timing. Agentic AI at scale started in mid-2025, and the multi-year erosion cycle takes years to play out. The firms running into this pattern are not announcing it. So I am making the structural argument without a current named case to point at, and the asymmetry between this pattern and the workforce-without-workflow pattern (Klarna) is itself part of the story. Workforce-without-workflow failures produce public reversals. Workflow-without-workforce failures are quieter, slower, and harder to see.

Inside the function

Decision-structure redesign and operational-process redesign play out at the firm-wide and process levels. There is a third move that gets less press and is doing real work in its own right: function-level redesign. A specific function inside the firm, often a knowledge function with high volume and high routine content, gets reshaped to run with agents doing most of the work and humans concentrating at exception, oversight, and orchestration. The Big Four professional services firms are the most public example of this third move.

Read these next three with one eye on a wrinkle. The firms selling agentic-AI transformation to enterprises (KPMG, EY, PwC, Deloitte) are simultaneously running it on themselves. When you read about what Big Four firms are doing inside their own audit, advisory, and consulting practices, you are reading both an industry shift and a case study from the firms that are now proposing similar redesigns to your enterprise. That dynamic does not invalidate what they are saying about their own practices. It does mean that the framing of what they are doing is shaped by what they are trying to sell.

Read the public statements from KPMG, EY, and PwC in sequence and you get three different framings of the same underlying play. KPMG talks about where the technology limit sits. EY talks about its own role as “client zero.” PwC talks about the retrofit-to-reimagine shift directly.

Thomas Mackenzie, KPMG’s audit chief digital officer, said in April 2026 that for routine audit testing, “I think there will be next to no human beings in that bubble. It will all be agents and orchestrators doing it.” The hedge (“I think”) and the scope qualifier (“routine audit testing,” not full audit) are both load-bearing and worth keeping. KPMG is staging this rollout across the rest of the decade, with the routine portion of audit testing going first and broader audit work retaining human review, data collection, and risk assessment.

Mackenzie was also willing to name the harder question out loud: “It sounds like we’re just slicing the bottom of the pyramid off,” he said. “We’re lifting the pyramid up.” Whether the pyramid is sliced or lifted is the kind of distinction that will play out in the labor data over the next several years; what is clear is that the bottom is changing shape, fast. Mackenzie also said in the interview that KPMG will not hire college graduates anymore to create workpapers. That role is going to agents.

Janet Truncale, EY’s Global Chair and CEO, framed her firm’s approach differently. Her April 2026 press release positioned EY as “client zero.” “As we help the world’s leading organizations transform,” she said, “we hold ourselves to the same standard, starting as client zero.” EY’s commitment is wider in scope than KPMG’s:

all of its roughly 130,000 Assurance professionals, all phases of the audit, full rollout by 2028. EY Canvas, the firm's audit platform, processes about 1.4 trillion journal lines per year, which is the technical infrastructure that makes that rollout plausible.

Matt Wood, then PwC's Global and US Commercial Technology and Innovation Officer (he has since moved to AWS), named the shift more directly than either of the others. In a December 2025 piece, he framed the firm's 2026 work this way: in 2025, organizations had fit AI around their existing workflows; in 2026, the work PwC would do was about "helping clients flip that model" and designing processes with AI in mind from the outset. Wood is restating retrofit-to-reimagine in the vocabulary of a Big Four practice leader.

Deloitte has been publicly more cautious. Will Bible, a senior Deloitte audit partner, has said in interviews that "audit quality ultimately depends on human judgment, professional skepticism, and accountability." Three firms leaning hard into the redesign, one publicly hedging (for now).

The pattern at function level shows up in other industries as well. Moderna, in the pharmaceutical industry, has converted significant portions of its HR function to operate through agentic systems. Tracey Franklin, Moderna's chief people and digital technology officer, described one part of the redesign this way: "It's like your virtual HR, AI agent. It's what would normally be a junior-level HR analyst type, we've now converted into a GPT." Moderna's HR-Tech function operates as a single integrated structure, with Franklin holding both the people and technology leadership roles after a 2024 merger that put both reporting lines under one executive.

How that merger is governed is itself part of the coupling argument. The executive structure that owns workforce decisions and the executive structure that owns technology decisions had become entangled enough that running them under separate leadership produced friction. The merger was the formalization of work Franklin and the previous CIO had already been doing together for years. What Moderna reimagined here is more than the path to running a function. It is the function itself, collapsed with what used to be IT into a new function organized around architecting the flow of work. That destination did not exist on Moderna's org chart a few years earlier.

When KPMG, EY, or PwC arrives at your firm with a transformation proposal, they will be drawing on the redesign they are running internally. The case study is real. It is also a sales document.

Direct about what's mixed

Three things to be direct about before we leave this.

First, the workforce-reduction news of the past year is a complicated mix that does not survive being flattened into a single "AI is replacing workers" headline. Mo Koyfman, a venture capitalist at Shine Capital, made the observation publicly that some portion of what is being attributed to AI in workforce-reduction announcements is overdue right-sizing using AI as the cover story. "Most companies, if not all, could cut thirty to fifty percent of their workforce at any time and see no material difference in performance," he said. "AI has given air cover, more importantly, to execute on the right-sizing that you probably needed to do a long time ago."

This is a different thing from workflow-and-workforce coupling. Some announced reductions are coupling done deliberately. Some are right-sizing dressed in AI vocabulary because that framing plays better with investors than "we hired too many people in 2021." Adjudicating which firms are which is beyond what I can do from public information. The point is that the public framing of workforce reductions is shaped as much by what makes for good investor communication as by what is operationally true.

Second, "fully agentic" is overselling what the cases actually show. Across function-level redesigns where the framing has been most aggressive, including the Big Four, what is happening operationally is not zero humans. It is a much smaller human team concentrated at orchestration, escalation, and oversight edges, with agents doing the routine volume. The function compresses dramatically. The human role compresses with it. The role does not vanish, and the cases that have promised it would have been quietly walking back the promise. In February 2026, Gartner predicted that 50 percent of companies that attributed workforce reductions to AI in customer service would rehire by 2027, with similar functions running under different job titles. The pattern of compression is real. The version of it that involves complete elimination of human work in a function is mostly aspirational.

Third, the personal stakes here are real, and I am signposting them rather than trying to answer them in this part of the book. The reader who has spent a career building and leading human teams may now be looking at a hybrid human-agent organization where the team is smaller, the agents are more capable, and the executive's own role is changing shape. Perhaps you will find your work more leveraged and more strategic. Perhaps you will find it less personally meaningful, because what made the work feel important was the leading-people part, and there are now fewer people to lead. Both reactions are valid.

The closing chapter of this book takes that question seriously. For now, what is worth registering is that workforce redesign is not just an operational question. It is also a question about what the executive job is becoming, and the answer is going to look different depending on who is asking it.

• • •

Key takeaways

- **Workflow and workforce are one integrated decision, not two.** Treating them as two is the structural mistake many enterprises are still making.

- **Reimagine takes two shapes.** Decision-structure redesign collapses the org chart (Block); operational-process redesign rewires the work inside the boxes (JPMorgan KYC).

- **Retrofit captures real value through both cost-out and speed-up.** The advantage tends to compete away because copying is easy.

- **At function-level redesign, the firms selling you the transformation are simultaneously running it on themselves.** Read their case studies as both descriptions and sales documents.

- **Some workforce-reduction announcements are deliberate coupled redesigns; some are right-sizing using AI as cover.** The framing matters less than what is actually happening inside the function.

• • •

This chapter has covered the first of three linked decisions (settings) for the boat, the one about its design and crew. If you have read this far, you know the case for reshaping is structural and the case for leaving it alone has a half-life. What you may not yet have is a clear view of what the next decision is going to demand of you.

The next decision is linked to this one, not stacked on top of it. As the boat is being redesigned and the crew configuration rethought, the question of who needs to see what arises in the same moment. An agent operating in a redesigned workflow needs information the old workflow did not require. New handoffs require new context. New decision points require new awareness of what is happening elsewhere in the operation. The crew that used to focus on the work directly in front of them now needs to see further. The amount and breadth of context you provide is what determines whether the redesigned work can actually be done. It is also, as the next chapter will argue, considerably more uncomfortable than it first appears.

That is Tension 2.

Further reading

Companion website: riverdoesntwait.com

A few readings to go deeper on this chapter. Sources covered in earlier chapters are not relisted here.

Microsoft 2026 Work Trend Index Annual Report

Microsoft surveys workers and managers globally every year on what is changing in the experience of work. The 2026 edition is one of the first to publish an empirical decomposition of where AI impact actually comes from at the enterprise level. The headline finding, the random forest result on roughly nineteen thousand workers showing organizational factors at about twice the impact strength of individual factors, is the strongest empirical anchor for the argument I am making here that the constraint is the operating system rather than the people inside it. Microsoft has commercial interest in the conclusion, since Microsoft sells the agentic platforms that the redesign relies on but the methodology is unusually transparent for a vendor-published study, and the finding is consistent with what the named-enterprise cases independently show.

The Block-Sequoia essay and the Dorsey podcast appearance, March and April 2026

Block CEO Jack Dorsey and Sequoia partner Roelof Botha co-published an essay titled "From Hierarchy to Intelligence" on March 31, 2026, framing Block's reorganization as building the company as an intelligence rather than as a hierarchy. Dorsey followed up on Sequoia's "Long Strange Trip" podcast on April 2, 2026, with the verbatim "in the most ideal case, there is no layer" framing and the specifics on current and target layer counts. Read both. The essay is the strategic framing; the podcast is the operational concrete. Block is the fullest public case of decision-structure redesign, with the founder doing the unusual work of being publicly explicit about both the play and its limits.

JPMorgan Company Update, February 23, 2026

JPMorgan's annual Company Update is dense with substantive material on what an at-scale agentic AI program looks like inside the largest US bank. The Dimon "we have displaced people from AI" framing in context (paired with redeployment infrastructure at scale and active plans to expand it further) is the strongest CEO-voice anchor available for what coupling done deliberately looks like. Erdoes on the AWM controls scale-out (200 to 3,000 to another 3,000-5,000 next) is the operational concrete behind the framing. Worth reading or watching in full rather than via press summaries; the press coverage tends to crop the framing in ways that lose the operational substance.

WSJ, "Why Moderna Merged Its Tech and HR Departments," May 12, 2025

The longest single primary source on the Moderna case, the one that plays out at the governance level - how the function is run and overseen rather than how its tasks get done. Tracey Franklin in her own words on the merger, the "virtual HR agent" framing, and the redesign of work across the firm based on what is best done by people versus what can be done by GPTs. Franklin's voice in this article is the most candid public CEO-voice anchor on the governance form of the coupling argument (the claim that workflow change and workforce change have to move together).

WSJ, "KPMG and the Future of Audit," April 10, 2026 (Mackenzie interview)

Thomas Mackenzie on KPMG's audit redesign, with the "I think next to no human beings" framing, the "slicing the bottom of the pyramid versus lifting the pyramid up" framing, and the operational specifics on routine audit testing as the first wedge. Pair this article with EY's April 7, 2026 press release on the 130,000-auditor agentic rollout and PwC's Matt Wood interview from December 31, 2025; reading the three together is the best way to see the buyer-seller dynamic at the function level.

Michael Hammer, "Reengineering Work: Don't Automate, Obliterate," Harvard Business Review, 1990

The historical anchor for the coupling argument I am making here. Hammer's article on Ford accounts-payable is more than thirty years old and reads even better now than it did when it was published, because the technology that finally makes its argument completely operational is the agentic AI of the past two years. The category-error framing is still load-bearing. Worth reading in full if you have any doubt that the discipline of coupled workflow-and-workforce redesign predates the current technology.

NBER Working Paper 34984, "Artificial Intelligence, Productivity, and the Workforce: Evidence from Corporate Executives," March 2026

A Federal Reserve Bank of Atlanta-led collaboration with Richmond Fed and Duke, surveying corporate executives on AI productivity and workforce effects. The headline finding (compositional reallocation of labor within and across firms, with routine clerical work exposed to substitution and analytical and managerial work more often complemented) is convergent with the "squeeze in the middle of the work" framing above at the broader level of work allocation. Worth reading directly rather than via the secondary commentary the paper has attracted; the paper is more careful than most of its summaries.

Chapter Five

Tension 2: What the Agent Must Know

Why context provision is the choice that decides what kind of organization you are running

As the work is being redesigned, the AI agents helping to staff it likely have a problem you have not yet (knowingly) made yours.

They do not know what success looks like. They do not know what the firm cares about this quarter. They do not know the unwritten rule about which customer-service exception gets escalated and which one gets resolved quietly. They do not know the policy that lives in the head of the senior controller who used to be three desks away. Some of this a human in the role would have picked up over years on the job. None of it is automatic for a team that has just been reconfigured around new seats, new tools, and new handoffs.

The decision about what to tell them is the second Tension. It is also, of the three Tensions, the most personally implicating, because the decision is not really about information. It is about how much of your own awareness, judgment, and influence you make explicit so that the teams of people and agents you have just put into the redesigned workflow can act on it.

Context is more than data

Most discussions of "data for agents" stop at data, narrowly understood. Spreadsheets the agent can read. Customer records it can look up. The document store it can search across. These are necessary. They are not what I mean by context.

Context is what the team operates with and against. It is the data, yes, but it is also the goals the team is working toward, the constraints on how it can get there, the institutional norms the firm operates by, the policies that govern the function, the implicit knowledge a human in that role would have absorbed without anyone having taught them explicitly, and the knowledge of the wider marketspace that shapes what a sensible decision looks like in this firm versus a different one. It is all the information the team needs to do useful work, plus all the information about what useful work means here.

Some agents and some teams operate against a tightly bounded slice of all that. This account. This transaction. This single step in this single workflow. Most agents in production today operate inside a context envelope like this, because narrow context is easier to provision, easier to audit, and easier to control. For some (especially retrofit) work, that envelope is exactly the right size. The task is bounded, the variation is small, the decisions are local, and giving the agent a wider view would buy you nothing and cost you precision.

Other agents and other teams operate with awareness across the function, sometimes across functions, sometimes across organizational boundaries to partners, customers, and suppliers. The wider view enables work the narrower view cannot. Work that depends on seeing how a decision in one place lands in another, or on weighing trade-offs only visible at the scope of the whole firm. These tend to feel more like reimagine scenarios.

Neither shape is the right shape in the abstract. The choice between them is a choice about what the specific work in front of you actually requires. Most enterprises are not making the choice deliberately. They have wound up with whatever context envelope their systems made easy, regardless of what the work actually needed. What I am arguing in this chapter is that the choice deserves to be made on purpose.

One more frame before moving on. Context setting is not a new activity for you, and it is not specific to AI. Senior executives have been setting context for the humans on their teams their entire careers. The same scopes I am about to walk through have always applied: what your function does, what the rest of the firm does, what the world outside the firm looks like. Context setting for an agent uses the same vocabulary, the same scopes, and the same political difficulty as context setting for a person.

What is changing in the agentic age is two things. First, who you are setting context for. The function in front of you increasingly contains both humans and agents working as a hybrid team, and the context you provide has to fit the system as a whole (connected to workforce and workflow redesign). Second, what the system can do with the context once it has it. Agents can hold a larger context in working memory than any human, operate on it at machine speed, and never get tired. That changes what becomes possible when the context is rich. It does not change what context is, or what your role in providing it is.

What you do and what the team needs are the same thing

There is a claim worth being clear about up front. It sounds strange the first time you hear it, then becomes obvious in retrospect.

Setting and controlling context is much of what a senior executive actually does. You decide what information flows to whom. You decide what outcomes the function is optimizing for. You decide what constraints apply to which decisions. You decide what your organization's goals and priorities are this quarter and this year and how those priorities get communicated down the line. You hold the institutional memory of why a particular policy exists and what it is meant to protect against. You bring in the external knowledge - the market, the regulatory environment, the competitor's last move - that makes a given decision sensible in this firm and not just in any firm.

That is most of your job.

Now look at what a team in front of you needs to operate well, whether the team is human, agent, or hybrid. It needs to know which information flows to whom. It needs to know what outcomes the work is optimizing for. It needs the constraints. It needs the priorities and how those translate to the work in front of it. It needs the institutional memory. It needs the external context.

The two lists are the same list. They have to be. From the executive's side, the activity is called "setting context." From the team's side, the activity is "having the context they need." Recent practitioner vocabulary, covering the agent-specific version of the activity, has settled on "context engineering." Different language for the same activity, depending which side of the question you are standing on. The scope of context an executive can set is the same scope as the executive's organizational reach, because the activities are the same activity.

That framing has an implication that matters for what this chapter is about and what it is not. Providing rich context to a team is not the same thing as transferring authority to that team. Authority comes from the combination of how the work is structured, what context is available to do the work against, and what oversight catches the team if it goes wrong. Those are the three Tensions of this book. Context provision sits alongside the other two, on equal footing. The three Tensions are not stages, and none is a precondition for the others. You work through all three at the same time as the work demands. Set context without workflow and oversight thought through and you have not actually delegated anything. Set workflow without context and the work cannot get done. Set oversight without context and you are governing a domain you have not yet defined.

The vocabulary worth pausing on for a moment is the practitioner vocabulary I just referenced. On June 18, 2025, Tobi Lütke, the CEO of Shopify, posted on X that he preferred the term context engineering to prompt engineering, because it described "the core skill better: the art of providing all the context for the task to be plausibly solvable by the LLM." A week later, on June 25, 2025, Andrej Karpathy replied to him in public agreement and offered what has since become the most-quoted formulation of the same idea: context engineering is "the delicate art and science of filling the context window with just the right information for the next step." On July 28, 2025, Gartner published a research note titled "Lead the Shift to Context Engineering as Prompt Engineering Fades." And on September 29, 2025, Anthropic's engineering team published a blog post titled "Effective context engineering for AI agents." Inside one quarter the vocabulary was established at technical altitude.

The technical altitude is not where I am working in this book. It is where AI engineers, platform vendors, and data-architecture teams are working. They are doing real and necessary work, and the vocabulary they have settled on is useful. What I want to do here, that they have not, is two things.

First, to elevate the conversation from technical altitude to executive altitude. The technical literature describes the activity at the level of tokens, retrieval mechanisms, and inference-time information assembly. That is correct as far as it goes. For the executive, the same activity is about what the firm is trying to do, who needs to know what, what trade-offs the firm is willing to make, and what information the firm is willing to share with whom. The technical vocabulary does not reach that far, and you cannot wait for it to.

Second, to name explicitly that context-setting for the team you are responsible for is what only a senior executive can do. Your CIO can architect the systems that hold and route the context. Your CIO can build the data layer, the integration plumbing, and the security perimeter. None of that is what I am pointing at. Business context, the kind I have just described, is what only the line executive can provide. Nobody else in the firm knows what the function is actually trying to do, and why, what the unwritten rules are, and what the firm is willing to trade against what. You are the only person in your organization who can provide it.

There is something extra challenging in this for many senior executives, though it is not the challenge of giving up authority. It is the challenge of having to make explicit what has been implicit. Most senior executives carry a great deal of context in their heads - priorities, constraints, institutional history, judgment about what matters and what does not - that they have never had to write down or articulate aloud. Setting context for the team in front of you forces that articulation. Things you knew in your head have to become things written down somewhere or said out loud. That confrontation between implicit and explicit is most of what this Tension is actually about.

Three levels of scope

The structural implication of all this is that the context required for any given piece of work sits at one of three levels relative to what you personally own.

Level one is context you own directly. Your function. Your team. Your decisions. Your priorities. The two activities (setting context as an executive and getting the team the context it needs) line up most cleanly here, because the scope of context is the scope of your formal authority. Most discussions of executive decision-making implicitly assume this is the only level there is.

Level two is context that sits inside the organization but outside your direct purview. Another function's processes. Another department's data. A peer executive's domain. Level two is where the most transformational reimagine work often lives, because the boundaries where two functions meet are exactly the boundaries the old workflow had to navigate by hand and the new workflow can route through differently. Your move at level two is to catalyze cross-functional coalition with peer executives whose scopes have to be combined for the work to make sense.

Level three is context that sits outside the organization entirely. Partners. Customers. Suppliers. The wider marketspace. The regulatory environment. Some of this may be owned internally by market-intelligence or partnership teams. Some of it may not be owned by anyone inside the firm and has to be acquired or negotiated for. Your move at level three is to catalyze cross-firm or cross-entity collaboration with external parties whose interests do not automatically align with the firm's.

Each level escalates the political and organizational difficulty of context provision. Level one is hard, but it is your hard. Level two is harder, because it requires getting other executives, with their own pressures and incentives, to combine context with you. Level three is harder still, because the external parties have their own organizations to run and their own reasons to share or to withhold context.

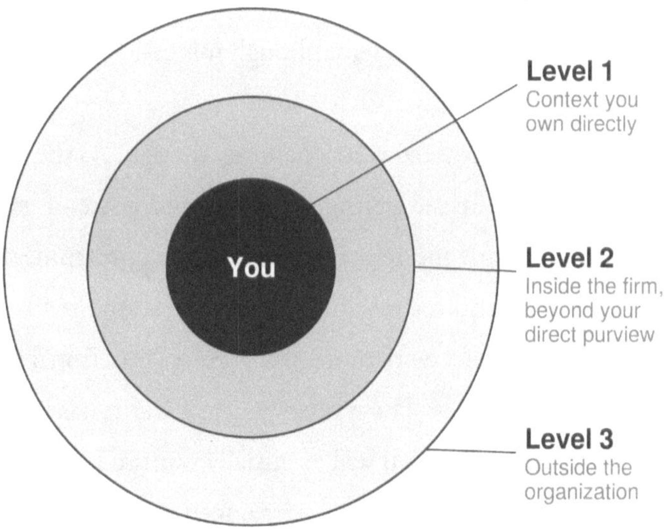

Choosing the appropriate level of context depends on what the work requires

A note on how the three levels relate to the retrofit-to-reimagine spectrum from earlier chapters. The levels are not stages you pass through as you move from retrofit toward reimagine. They are not steps and not a progression. Think of the diagram you just looked at as a target, with you and your function at the center. For any given piece of work, the question is which ring on the target you have to reach to assemble the context the work actually needs. Some work, including some genuinely reimagined work, lives entirely at level one (inside what you already own) and the context architecture you build for it never has to leave that ring. Other work cannot be done without reaching into level two or level three, no matter how modest your ambition for it. Where you aim on the target is set by what the work requires (closer to the center is not necessarily better, although it is easier/quicker to manage).

There is a statistical pattern worth naming, though. As the current in the river you are navigating speeds up and more of your function moves toward the reimagine end of the spectrum, the share of work that requires reaching beyond level one tends to grow, because reimagined work more often crosses functional and organizational boundaries. The tendency is real and it has implications for what coalitions you will eventually want to have built. But the tendency is not a rule. Some reimagined work stays inside level one. Some retrofit work touches level two. Treat the levels as scopes to aim at, each chosen for the work in front of you.

A second note on what "executive scope" really means here, because you will need it for the cases that follow. Your scope is not just your formal authority. It is the totality of your reach across the firm and beyond. That includes the peers you have coalition capital with, the executives you can get into a room and align with, the external network you bring with you, the market knowledge you have accumulated over a career. The gap between your formal authority and your actual reach is where your executive skill has always lived. In the agentic age, the same gap is where your additional context-provision capacity also lives. The wider your actual reach, the wider the context you can set. The scope of context you can set lines up with your full organizational reach, which extends well past your formal authority.

The empirical record across the three context levels

Take a look at where firms actually are along this structure. The empirical record at this point in the agentic-AI wave is consistent with what the three-level structure predicts. Level-one cases are widespread. Level-two cases exist, but they cluster at a small number of named lighthouse firms you have probably already heard about. Level-three cases are nascent, and the firms doing them are direct that they are early.

At level two, the visible examples I will highlight are at Moderna, ServiceNow, JPMorgan, and Advocate Health.

At Moderna, mentioned earlier, Tracey Franklin holds both the Chief People and Digital Technology Officer titles in a single role, formalized in late 2024 after a merger of HR and IT under one executive. The merger was not new work. Franklin and the prior CIO had been operating together for years before the formalization. The merger named what was already happening. Franklin's own framing, in her UNLEASH interview from June 2025, is direct: "This isn't just about HR or Digital, it's about rethinking, system-wide, how we build and rebuild the company to scale." Her articulation of the operational purpose is sharper still: "Together, we can truly architect the flow of work, how tasks, information and decisions get done." The combined function now operates as the level-two coalition unit for the HR-plus-technology axis of the company.

A different version of the same move shows up at ServiceNow. Jacqui Canney holds the title Chief People and AI Enablement Officer, announced in January 2025. Canney has been direct about why the coalition has to exist in the first place: "It can't just be like, here's the AI team, here's the employee workflow team, here's the CRM team. It has to literally go across." An operational detail from the ServiceNow HR team illustrates what the coalition does in practice. The team developed more than a thousand candidate use cases for AI inside the function, then built a rubric to evaluate them and refined the working set down to twenty-seven (per Josh Bersin's November 18, 2025 podcast interview with Canney).

At JPMorgan the level-two coalition is unusually visible. Mary Erdoes (CEO of Asset and Wealth Management), Lori Beer (CIO), Teresa Heitsenrether (CDAO), and Robin Leopold (head of HR) are named in public company materials as operating-committee co-sponsors of the bank's agentic-AI program. Most firms doing this work do not name their coalitions in their earnings calls.

A health-system version of the same move comes from Advocate Health, where the joint Chief Digital and AI Officer is Andy Crowder. His framing in the May 14, 2026 alliance press release captures the level-two move directly: "Our collaboration with Anthropic and PwC isn't about deploying technology for its own sake - it's about building the foundation that allows our 167,000 teammates to do more for every patient, in every community we serve, including the rural communities that need us most."

Level three is more nascent. The publicly named operational cases include PepsiCo with Siemens and NVIDIA, and TraceLink in pharmaceutical supply chain.

At CES 2026 on January 6, PepsiCo, Siemens, and NVIDIA announced a multi-firm digital-twin collaboration. The architecture is a shared "physics-level" model of PepsiCo's production facilities, with AI agents simulating, testing, and refining proposed system changes against the shared model. The joint release reports identification of up to 90 percent of potential issues before any physical modifications occur, a 20 percent increase in throughput on initial deployment, and 10 to 15 percent reductions in capital expenditure. The structural point holds independent of the precise numbers: PepsiCo, Siemens, and NVIDIA are operating against a shared context model that none of them individually owns and that none of them could have built alone.

TraceLink, the pharmaceutical supply-chain platform, announced on February 28, 2026 an expansion of its agentic orchestration across what it reports as a network of more than 310,000 trading partners (manufacturers, distributors, pharmacies, wholesalers). The architecture, in TraceLink's framing, is multi-firm context sharing as a precondition for cross-firm agentic coordination on shipment tracking, exception handling, and serialization compliance. The level-three move at TraceLink is structurally different from PepsiCo-Siemens-NVIDIA. Instead of a small number of firms sharing context for a defined project, TraceLink has built infrastructure for many firms to share context with each other through TraceLink's platform.

Two publicly named operational cases at level three, in two different industries. The frontier exists. It is small (so far). But the structural argument about why level three matters and how it relates to the retrofit-to-reimagine spectrum does not require dozens of cases. It requires one or two that show the move is real, which the public record now supplies, and an acknowledgment that the rest of the field has not caught up, which is also true.

A journey, not a deliverable

Three levels of context scope, no specific one required for any particular piece of work, but each of them harder to reach into than the one before. You may be looking at this and concluding that the level-two and level-three demands are beyond your reach, which is exactly the conclusion that reduces your ability to reimagine.

I'll handle that directly, because the alternative is to leave you somewhere I do not intend to.

The context architecture for any specific function is something you build out over time. You do not need it complete on day one, and you do not need it to reach into every ring of the target. You start with the context the work in front of you requires. You build the architecture as the work surfaces more. You reach into level two when a specific piece of work calls for it, and into level three when one calls for it, and not before. For some pieces of work, you will never need to reach beyond level one, and that will be the right answer for those pieces of work. The journey is the buildout of context that the work in front of you actually demands.

Moderna's own deployment captures the shape concretely. As of December 31, 2024, Moderna reported sixteen hundred custom in-house AI tools in use across the company (per the 2024 Annual Report). Six months later, the figure had grown to more than three thousand. The architecture was being built out as the work surfaced what was missing, function by function. Nobody at Moderna sat down on day one and designed three thousand of them. They built the first batch and then they kept going.

The ServiceNow trajectory from a thousand candidate use cases down to twenty-seven is the same kind of artifact at a different scale. The level-two coalition at ServiceNow did not arrive at twenty-seven by guessing. It built a rubric, evaluated, prioritized, and selected. The journey at level two has its own discipline. The discipline is what produces a coherent architecture rather than a thousand half-implemented experiments.

The journey framing is what protects me from leaving you in the loss-of-hope outcome the three-level structure otherwise risks.

You do not have to have it all figured out. You have to start where you are, build what is in front of you, and recognize that the right move at a higher level next quarter is a different move from the one you make today.

Architecture, not plumbing

If a context envelope is something someone has to design rather than something that happens automatically, what is the design work? The clearest public framing comes from Jack Dorsey and Roelof Botha, in the essay I referenced in earlier chapters, where Block is described as building a "company world model" and a "customer world model" with agents operating against those models. The essay introduces additional vocabulary: an "intelligence layer" between the work and the people, "people at the edge" who own outcomes rather than supervising hierarchies. The vocabulary is the sharpest articulation in the current public record of what context-as-architecture means at the scale of an entire company.

The operational substance is carried by cases inside specific functions. Three of them follow, at three different scales.

Moderna's HR architecture - a single function

Moderna's HR-Tech function has built a deliberate context architecture rather than connecting agents to a shared data lake and hoping something would emerge. The architecture has an entry-point assistant ("Ask HR") that handles routine inquiries and routes complex cases to specialized sub-assistants covering performance management, equity compensation, benefits administration, and several other domains. Each sub-assistant operates against its own bounded context: domain-specific policies, escalation rules, decision criteria, and the data slices relevant to its function. The architecture grew organically as the work demanded. Hem Patel, a senior HR executive at Moderna, described at the WorldatWork Total Rewards conference in May 2025 that the self-review assistant is the most-used internal tool inside the function, with the US benefits assistant second. Franklin's framing of the function-executive role captures the work concretely: architecting the flow of work, not piping data to it.

A version that spans a cross-functional process, rather than sitting inside one function, looks structurally similar.

JPMorgan's AWM controls architecture - a cross-functional process

JPMorgan's controls function inside Asset and Wealth Management built an architecture of about eighty specific prompts, each scoped to a specific control task, each operating against the relevant policy-aware data, each producing auditable output. The architecture was not a generic chatbot pointed at a data warehouse. It was a deliberate decomposition of the controls work into prompt-specific contexts, each context curated for what the agent and the humans alongside it needed in order to do that specific task well. The trajectory of the controls function (two hundred people in the pilot, three thousand currently doing the redesigned work, another three thousand to five thousand identified for the same treatment) is itself a journey artifact. The eighty prompts were not a one-time design. They are themselves a context architecture that keeps growing as the work surfaces what is missing.

MetroHealth pharmacy - a single operating unit

To broaden out from financial services and pharma, MetroHealth's pharmacy department deployed an agentic system specifically for medication prior authorization (PA). The system operates against a context of payer-specific medication rules, formulary data, patient clinical context, and a decision-support workflow built by the pharmacy team. Cycle time on medication PA dropped from approximately eighteen hours to about five minutes, with a 15 percent increase in approval rates, per Becker's Hospital Review reporting on the deployment in May 2025. The number itself matters less than the structure. The pharmacy team built the context architecture for one specific kind of decision and let the agent operate against it.

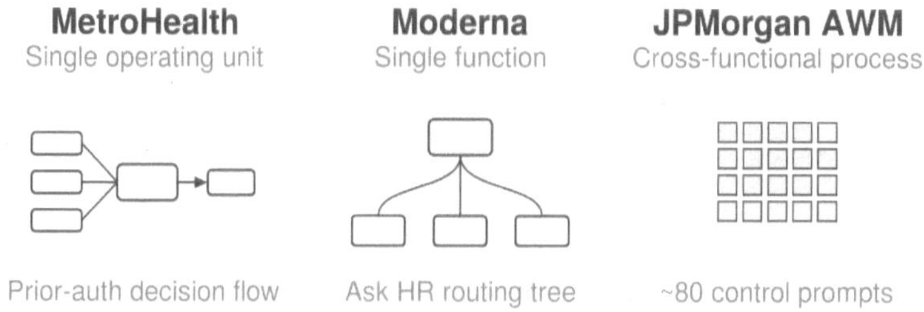

Three context architectures. No single approach is best for all situations

Three cases at three different scales, plus Block as the whole-company framing. What is common across them is that the team is operating against context someone deliberately designed. None of it happened to be there by accident. The shape of the context is part of the design of the work.

The architecture framing in an enterprise CIO's own voice comes from Lori Beer, JPMorgan's CIO, in her Fortune interview from April 29, 2026. Beer's metaphor for the firm's broader software-development redesign is the factory. The work, in her telling, is rebuilding the factory of how software gets made at scale: moving beyond the integrated development environment, giving agents the context they need to handle complex tasks, with senior engineers reviewing the agent output rather than writing the code themselves. Same point in different vocabulary. The context architecture is what is being rebuilt. The factory is what a context architecture looks like at production scale.

Outcome setting is the highest form of context

Now the part of context setting that sounds suspiciously simple.

The higher in the leadership hierarchy you go, the more of the context you provide concerns outcomes relative to informational or procedural data. A frontline manager provides context that is mostly procedural: do this step, then this step, then this step, based on this data. A senior executive provides context that is mostly outcome-based: the function needs to ship this by this date, with these constraints, optimizing for these things over these other things. A CEO provides context that is almost entirely outcome-based and only barely operational: this is what the firm is for, this is what we will and will not do, this is what we are trying to become.

Context provision at the executive level is less about plumbing and more about ensuring a shared understanding with the team regarding what success looks like. That is what the senior executive has always done with humans. It is what you now have to do, more explicitly than before, with hybrid teams including AI.

A few executive voices in the public record make the point concrete.

Lori Beer, at JPMorgan, on what the firm has been working on at the agent-design level: "We have been focusing very early on, on simple things, like what's the right level to create an agent; how do you give them identity and access?" The framing in Fortune's reporting of the interview describes Beer's permissions architecture as one where agents are scoped narrower than the humans whose tasks they perform. A deliberate decision that the agent gets a tighter envelope than the human did. Beer's own framing of why: "You don't want them to go outside the bounds of the specific tasks that they can do, because they don't have the same thinking a human does." The right level to create an agent is, in Beer's working vocabulary, a question about outcomes and bounded permissions, not procedures.

Tracey Franklin, at Moderna, in the UNLEASH interview already cited: "This isn't just about HR or Digital, it's about rethinking, system-wide, how we build and rebuild the company to scale." That is an outcome statement covering a whole function. The level of context Franklin is providing is not "here are the workflows we are redesigning." It is "here is what the redesign is for and what we are trying to become." The HR-Tech function then works out the procedural specifics in service of that outcome.

Jack Dorsey, in the April 2026 Sequoia podcast appearance, on what executive work becomes in an agentic organization: judgment and alignment, not procedural specification. Dorsey's stated ideal of zero management layers between the CEO and the individual contributor is, structurally, the outcome of having pushed procedural context down to the team (whether human, agent, or hybrid) and concentrated human effort at two points: doing the work, and making the judgment calls that set its limits.

That is not the only valid shape for context provision. An agent doing a tightly bounded task against tightly specified procedures is operating against a different shape of context entirely, and for that task the procedural shape is exactly right. The outcome-based shape and the procedural shape are different tools for different work, each right for its own kind of task. The shape you reach for depends on what kind of work you are setting context for and what kind of latitude you intend the team to have in doing it.

Direct about what's mixed

On the diagnostic question of why agentic deployments fail. I have not, in this chapter, walked through the literature on agents that failed because they were not given the right context. The literature does not really exist yet. Public failure postmortems for specific under-contextualized deployments are thin, partly because the firms running into those failures are not announcing them, and partly because the diagnostic itself is hard.

When an agent does not deliver the expected value, the cause might be a capability gap. It might be a scope mismatch in the workflow redesign. It might be a workforce-alignment problem. It might be a governance gap. Or it might be that the agent was under-contextualized for the work it was trying to do. The diagnostic is entangled across the three Tensions, which is itself one of the reasons this book treats the three Tensions as needing to be navigated in unison rather than diagnosed in isolation.

The strongest empirical evidence for the under-contextualization claim comes from the inverse. The firms that have built deliberate context architectures (Moderna, JPMorgan, MetroHealth on pharmacy PA) reveal, by what they have explicitly provided, what less successful deployments have been operating without. Not the technical specifics that would live in the heads of system architects, but the more fundamental things: clarity on what the team is actually trying to accomplish, awareness of what the rest of the function is doing, knowledge of the relevant constraints, and access to the institutional memory of why this firm does this kind of work the way it does. The lighthouse cases reveal the missing items by showing what they made present.

Context and control travel together

Before turning to the next chapter, I want to name where context provision and bounded authority already show up in the public record as inseparable, because the empirical pattern itself is the bridge.

Lori Beer's Fortune interview, already cited several times above, is one place this shows up explicitly. Beer's framing of the right level to create an agent is a single decision that joins context and permissions. Agents are scoped narrower than the humans whose tasks they perform. They are given the data they need and excluded from data they do not need. The context provision and the permissions architecture are designed together because, in Beer's working frame, they have to be.

JPMorgan's own technology blog from March 23, 2026 makes the same point explicitly. The post, titled "Securing the next generation of AI agents," names three properties that compound dangerously when an agent has all of them at once: processing untrusted inputs, accessing sensitive data, and authority to act externally. The blog calls that combination "the lethal trifecta." The piece names persistent context, system memory, and external content as vectors through which agent behavior can be manipulated when authority and runtime controls do not travel alongside the context provision itself.

The recommended posture is what the blog calls "policy-aware data at runtime": data that travels with labels about what it is, how sensitive it is, and what it is allowed to be used for, with the rules enforced at the point of execution, producing an auditable trail. JPMorgan's summary line is the one worth carrying into the next chapter: "align safeguards to capability and risk." We will discuss this further in the next chapter, when we address built-in governance.

The technical detail is not the takeaway. The structural point is. Context provision and bounded authority travel together at the reimagine end of the spectrum. They are not sequential decisions you can settle in order, finishing context this quarter and starting autonomy next. They are joint decisions made in unison, with the joint outcome being whether the team operates safely against rich context or unsafely against the same.

This is the empirical version of the foundational premise of this book. The three Tensions navigate in unison along the retrofit-to-reimagine spectrum, with each Tension having its own internal journey of buildout. Within Tension 2 specifically, the unison shows up empirically in named cases. The Beer and JPMC framings are primary-source evidence that practitioners running large-scale agentic deployments treat context and bounded authority as inseparable. They treat them that way because the work, in their setting, does not work otherwise.

• • •

Key takeaways

- **Context is more than data.** How much to give and at what scope is an organizational decision, not a technical one, and it deserves to be made deliberately.
- **The scope of context you can set is the scope of your organizational reach.** What an executive does and what a team needs to operate well are the same activity, seen from two angles.
- **Providing rich context is not, in itself, a transfer of authority.** Authority comes from context, workflow design, and oversight navigated together; none of the three by itself is enough.
- **Three levels of context scope: what you personally own, what sits inside the firm but outside your purview, what sits outside the firm entirely.** Appropriate scope depends on what the work requires, not a sequence to march through.
- **Context is architecture, not plumbing.** The architec-

ture gets built out function by function as the work in front of you demands more, not as a deliverable to finish.

• • •

The boat now has people and agents in redesigned seats, and a deliberate context architecture being built around them. The next question is the one Beer and the JPMC technology blog have already been answering inside their own organization. How much can the team do without checking back with a human? And what catches a bad call before it becomes a bad outcome?

The two questions are joined to each other, and they are joined to what you have just decided about context. Rich context, especially outcome-based context, enables wider autonomy without recklessness. Narrow context forces narrower autonomy, because the team does not have what it would need to act safely beyond a tight envelope. Bringing the three Tensions into alignment is the actual work of running an organization in the agentic age.

That is Tension 3.

Further reading

Companion website: riverdoesntwait.com

A few readings to go deeper on this chapter. Sources covered in earlier chapters are not relisted here.

Fortune, "How JPMorgan's CIO is reshaping work at the bank with a $19.8 billion annual tech and AI budget" (Lori Beer interview, John Kell, April 29, 2026)

Beer is the clearest senior-CIO voice in the public record on how agentic context provision and bounded authority are being designed together inside a large bank. The "right level to create an agent" framing, the "rebuilding the factory" metaphor for software-development redesign, and the permissions-architecture detail are all in the same piece. Worth reading in full rather than via press summaries.

JPMorgan Chase Technology Blog, "Securing the next generation of AI agents" (March 23, 2026)

The bank's own technical articulation of why context provision and bounded authority travel together at the reimagine end. The "lethal trifecta" framing on agents that combine untrusted inputs, sensitive data access, and external action authority is the most precise vocabulary in the current public record for the configuration that requires the highest level of continuous runtime safeguards. The piece is technical and brief; read it for the structural framing rather than for the implementation specifics.

UNLEASH, "Why Moderna merged HR and IT to better 'architect the flow of work'" (Tracey Franklin interview, Allie Nawrat, June 27, 2025)

Franklin's own framing of the HR-Tech merger, the "Ask HR" routing architecture, and the system-wide rethinking. The architect-the-flow-of-work language is in her voice. A strong primary source on what level-two coalition looks like from the executive holding the joint title.

McKinsey, "The State of AI" (QuantumBlack, November 17, 2025)

The most-cited large-sample survey on the state of agentic AI adoption in mid-to-late 2025. The decomposition matters: 62 percent of organizations at least experimenting with AI agents, 23 percent scaling somewhere in the enterprise, no single function above approximately 10 percent reporting fully scaled agents. Useful corrective to coverage that has conflated the experimenting figure with the scaling figure.

Andrej Karpathy on X, "+1 for 'context engineering' over 'prompt engineering'" (June 25, 2025) and the preceding Tobi Lütke post (June 18, 2025)

The two X posts that started the practitioner-vocabulary cluster. Lütke's preceding post is the originator. Karpathy's "delicate art and science" formulation in reply has become the most-cited verbatim across the secondary literature. Reading them in order is the most direct way to see the conversational arc that the analyst firms later joined.

Chapter Six

Tension 3: Built-In or Bolted-On

Why the governance you have today was built for a different shape of work

In November 2022, Jake Moffatt's grandmother died. He went to Air Canada's website to book a flight to the funeral. While he was there, he asked the airline's chatbot whether the bereavement-fare discount, which Air Canada does offer, could be applied to a full-fare ticket after the fact. The chatbot told him yes.

The actual policy, on a page of the same website, said no.

When Moffatt tried to claim the discount, Air Canada refused. He took the airline to a tribunal in British Columbia. Air Canada's defense was that the chatbot was a separate legal entity, responsible for its own statements. In February 2024, the tribunal called that submission remarkable, and ruled for Moffatt.

The damages were $650. The interesting thing in the case is not the money. It is the boat that produced the conversation.

The boat, in this case, is Air Canada's customer service organization. Trained agents on the phones. Written policies that bind what those agents can say. A website with thousands of pages, written by content teams, reviewed by legal, kept in sync with the policies. Sometime in 2022 or 2023, the organization put a chatbot on the website. The chatbot was not reviewed the way the rest of the operation was reviewed. Nobody asked, before it launched, whether what the chatbot would say about bereavement fares matched what the page said, or what a trained agent on the phone would say. Until Moffatt asked it on the wrong day, nobody had to.

This chapter is about that gap. This is Tension 3: how much room a piece of work has to run on its own, and what catches a bad call when it does. There are two ways to keep a boat off the rocks. You can post a lookout, ring the bell when the rocks come into view, and turn hard. Or you can know the chart, set the course around the rocks, and never need the bell. Both ways can work. Both have always worked. Which system you run, for which kind of work, is the question.

Two ways to govern the boat

Three words come up over and over when people in this field talk about agentic AI: supervision, autonomy, governance. **Supervision** is what a reviewer does, moment to moment, when the team is acting. The reviewer can be a human, but in many newer architectures it's another agent watching one level up from the agents being supervised. **Autonomy** is how much the team gets to do without that supervision. **Governance** is the structural arrangement that decides what supervision and autonomy mean here. Three views on a single question. How does work get done well when the work could go wrong?

There are two answers, and they sit at opposite ends of a spectrum.

Bolt-on governance is the boat's lookout. The team does the work. Reviewers sit alongside the team, watching for things that look like trouble. Compliance, audit, legal review, model-risk committees, responsible-AI boards. The arrangement is built around catching problems before they go live. When something gets through, you fix it, you patch the policy, you carry on.

Built-in governance is the boat's chart. The team's operating context already encodes the rules. The policies that say what is and isn't allowed are part of what the team operates against, the same way the goals and constraints are. The team's default behavior is the safe behavior, because the team's training, tools, and workflow have been set up so the obvious wrong moves are not available, the implicit wrong moves are off the menu, and the work the team produces is in bounds because the team has no path to producing it out of bounds. You still need a lookout occasionally. With built-in governance, the lookout is the exception. With bolt-on governance, the lookout is the operating model.

If those sound like the same thing said two different ways, bear with me for another minute. Bolt-on is a triage operation. Problem occurs. Problem caught, hopefully. Problem fixed. Built-in is a prevention operation. Problem does not occur, because the context the team is acting in does not lead the team there. You could do a really good job of catching errors. Or you could do a really good job of not making them. They are not the same achievement.

The names come from cybersecurity. Secure-by-design versus bolt-on security has been the framing in that field since the 1990s. The US Cybersecurity and Infrastructure Security Agency put out its "Secure by Design" principles in 2023. Bain & Company, in an April 2026 piece on agentic AI platforms, said it cleanly for the AI version: security and governance must be embedded by design, not bolted on after deployment. The vocabulary is older than the AI conversation.

There is a useful name for the gap that opens up when a bolt-on arrangement keeps running while the work underneath it changes. **Coherence debt**: the growing divergence between what a system is supposed to do and what it is actually doing in real time. The term comes from the Mindful AI Foundation, and you have seen it before: it is the gap Rao Mikkilineni's Foreword opened this book with. The Foundation's work on AI architecture is in the further reading. A bolt-on architecture, deployed in a world where the work has quietly changed shape, accumulates coherence debt the way a building accumulates structural strain. For a while, nothing happens. Then something does.

You already met built-in governance under a different name in Chapter 5. The whole argument of Chapter 5 was that the executive's job is to provide rich (enough) context to the team. What the team is trying to do. What trade-offs are acceptable. What the firm's constraints actually are. Rich context turns out to be the substrate that built-in governance operates against. The policies that say what is and isn't allowed are part of that context, not a separate layer bolted on top of it. They do their work because the team can see where its actions sit relative to them.

Obviously, this isn't rocket science. Setting proper goals and clear boundaries to operate within is the job of any great leader or manager, no matter whether the team is made up of AI agents, human beings, or some hybrid combination.

Context and governance are not the same thing. Context is what the team uses to act. Governance is what sets the limits on what the team can do. The two are tightly linked because, without rich context, built-in governance has nothing to bind against. With rich context, the rules built into the team's operating substrate can shape behavior from the inside rather than from outside.

When the lookout misses the rocks

The Air Canada case is a small one. The damages were $650. Two larger ones in the same window of time show the same pattern at higher consequence.

In June 2025, Microsoft quietly patched a flaw in Microsoft 365 Copilot. The flaw worked like this. An attacker sends a user an email. The user does not have to open the email or click anything in it. The Copilot, doing its normal job of helping the user, reads the user's inbox as part of building its context. The attacker's email arrives in that context the same way the user's own messages do.

Hidden in the body of the email are instructions: send the user's sensitive data to an outside address. The Copilot follows those instructions because it cannot tell, from inside its own context, that some of the words came from the user and some came from an attacker. To Copilot, all the words looked the same.

The researchers who found the flaw called it the first zero-click attack of its kind against a production AI assistant. It's zero-click because nobody had to actually click a malicious link to fire off the action. Microsoft's fix was not a filter. The company had to change how Copilot itself was built. How it parsed instructions. How it kept the user's content separate from outside content. How it decided what it was allowed to do with what it had read. The fix lived inside the assistant, in how it was built, not in the perimeter around it.

The bolt-on response, if there had been one, would have been to add a reviewer to check what Copilot was about to send before it went anywhere. That response would have caught some attacks. It would have missed others. The built-in response was to redesign the assistant so the rocks would not be in its course down the river in the first place.

Across the Pacific in the same window, the Australian Department of Employment and Workplace Relations published a 237-page independent review of its welfare-compliance IT system. Deloitte wrote the review under a A$440,000 contract. A researcher at the University of Sydney noticed something odd in the footnotes within weeks of publication. Academic papers were cited that did not exist. A federal court judgment was quoted that contained no such quote. More than a dozen references were fabricated. The revised version of the report, published a few weeks later, disclosed what the original had not: GPT-4o had been used to help produce the report. Deloitte returned the final payment of the contract.

The press read the case as one about AI making things up. It was. But the deeper read is about what the review machinery could and couldn't see. Deloitte's quality review process was built for reports assembled by analysts who had read what they were citing. The new kind of report still went through the same review. The same questions got asked. Is this well written? Does it cover what the client asked for? Does the analysis hold together? Yes, yes, yes. Are these citations real? Nobody thought to ask. In the world the review process came from, those questions caught the failures that mattered. They no longer do.

That's the pattern in all three cases. Air Canada's review process was built for static pages. The chatbot was not a static page. Microsoft's safety machinery was built for software that does what the user tells it. The agent took instructions from email it had read. Deloitte's quality review was built for analysts writing reports. The report was written by a system that could fabricate citations confidently and consistently, which a human analyst would never do. None of the three failures is a story about an agent being too autonomous. The envelopes the agents operated in were the right size. The thing watching the envelopes was the wrong shape.

Notice one more thing about the three. None of them is the full agentic article by the definition Chapter 1 drew. A chatbot answering a single question, an assistant misled while reading an inbox, a model drafting a report for human review: the first and the last sit below that line, and the middle one barely crosses it. That should not be reassuring. The review machinery in these cases missed failures from systems that had not yet started acting continuously, at machine speed, across workflows nobody watches end to end. The systems now arriving have.

This is what bolt-on looks like when the work changes underneath it. The lookout is still scanning the horizon. The rocks are no longer on the horizon. They are right in front of the bow.

The approach follows the work

The right approach follows the work, and the work sits on the same retrofit-to-reimagine spectrum that the rest of this book runs on. Bolt-on fits better at the retrofit end, because the inherited rules-and-audits arrangement was built for behavior the new system still produces in recognizable shapes. The reviewers know what to look at. The escalation paths catch what they were trained to catch.

Built-in becomes necessary as the work moves toward the reimagine end, because the inherited arrangement was not designed to catch behavior the system is now producing in shapes nobody was looking for. Mismatched approaches fail predictably. Bolt-on governance on reimagined work misses what it cannot anticipate. Built-in governance for retrofit work can be over-engineering or, at worst, unnecessary friction.

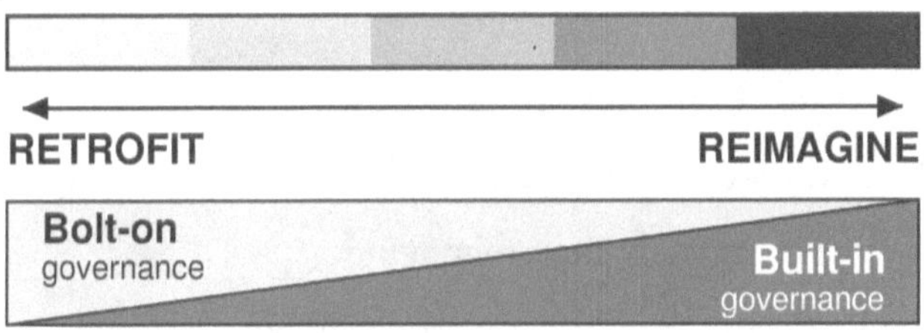

Built-in governance is more likely to be required the closer you are to the reimagine end of the spectrum

Most enterprises run both approaches across different parts of their boat. The approach is set deployment by deployment.

JPMorgan is the clearest demonstration of this in the public record. The bank built an internal LLM platform that at least 200,000 of its employees now use. The platform's design is described, by Lori Beer, the firm's CIO, in straightforward terms. Each agent is given a tight identity. The data the agent can see is bounded. The actions the agent can take are scoped more narrowly than those of the human whose work the agent is helping with. The substrate the agent operates against carries policy labels, enforced as the agent works. That is built-in.

Same interview, Beer also describes the firm monitoring the outputs the large language models are producing. That is bolt-on. The substrate prevents the dangerous combinations from forming in the first place. The output monitoring catches what the substrate did not prevent. The two layers are an explicit design choice. The reimagined parts of the bank run built-in. The retrofit-shaped parts run bolt-on. The firm runs both at the same time because the work runs both at the same time.

A few other examples worth knowing about, each leaning built-in for the work being described because most examples today are still closer to retrofit than reimagine.

When PwC launched its client-engagement platform PwC One in March 2026, the firm's then-Global and US Commercial Technology and Innovation Officer, Matt Wood, put it this way on LinkedIn: "This isn't AI layered onto existing processes. It's AI at the center of how an engagement works ... built-in transparency, verification, and expert oversight." Wood has since moved to AWS, but the launch framing was the point. The platform carries the methodology. The methodology carries the controls. Clients work inside the platform. The boat was redesigned, not retrofitted.

KPMG made the same move at roughly the same time with Workbench, its client-delivery platform. The press release describes Workbench as the foundation underneath KPMG firms' client delivery. The platform is built-in for client engagements. The firm's separately-run Trusted AI Framework is the bolt-on layer at the AI-governance level. Same firm. Different approaches. Different work.

Siemens runs built-in at the operational-technology layer of its industrial customers, because anything else is impossible there. A few months before the PwC and KPMG launches, Michael Metzler, who runs cybersecurity for Siemens' digital-industries business, described the firm's approach in a trade-press interview: defense in depth, every connected component required to verify and authorize the others, and the AI agents inheriting the same layered industrial-security architecture that every other component on a factory floor already runs under. In an environment where a wrong industrial control decision plays out in milliseconds, the lookout is not fast enough. Built-in governance is the only option.

Bob Pisani, the CTO at Addepar, the wealth-management platform, was on a podcast in May 2026 making the same kind of move from a different industry. The frame he used: AI built natively into the platform, not as a feature, but as infrastructure. The platform carries the architecture. The architecture carries the controls. Different industry. Same picture.

Once you can hear the difference, you can read the public record more accurately. Executives describing the built-in side of their work reach for structural words. Foundation. Infrastructure. The right level. Defense in depth. Architecture. These words name how the operation is built. Executives describing the bolt-on side reach for process words. Council. Framework. Closed-loop. Human in the loop. These words name what people do.

Sandeep Dadlani at UnitedHealth, talking to Fortune about how the company handles AI in claims work: AI is never used to deny a claim. If a claim is not eligible to be approved, it goes up to a human agent. That's the bolt-on lookout, by design, for work where the bolt-on lookout is the appropriate posture.

Himanshu Polavarapu at Verizon, on agentic deployment at the firm: Verizon sets up an AI Council, runs a responsible-AI framework, uses a closed-loop governance model that runs a proof of concept first and then deploys at scale. Process. Bolt-on.

Lakshmi Raman, the CIA's chief AI officer, on the agency's approach to deployments like help desks and form filling: always with a human at the crucial step to review what's going on. Same posture, same vocabulary tells.

None of those executives is describing the wrong arrangement. Each is describing which approach fits the work in front of them. The lesson is to listen for what an executive is describing right now, in this deployment, and to recognize which approach fits the work being described.

One more voice worth bringing forward, because it ties Chapter 6 back to Chapter 5 in a way nothing else in the public record does. Derek Waldron, JPMorgan's Chief Analytics Officer, was interviewed by McKinsey in October 2025 about how generative AI was changing roles at the firm. Waldron talked about a new job category emerging at the bank. The prompt engineer, he said, is evolving into what they call a context engineer, getting all the context needed into an AI system so it can make the right decisions.

Waldron was effectively making a governance claim. The decisions that are right include the decisions that are correct, yes, but also the decisions that are within bounds. The context engineer's job, done well, is what makes built-in governance possible. The reading I'm putting on Waldron's language is mine, not his, but the language does the work. Context is not governance but built-in governance requires rich context.

The current is faster now

None of this is new.

The bolt-on / built-in choice predates the AI conversation by decades. Cybersecurity has been running the same debate, with different vocabulary, since the 1990s. The legal profession has its own version, called regulation by design or compliance by design. Aviation has been doing built-in since the cockpit-discipline reforms of the 1980s. Surgical suites have been doing it since the time-out and safety-checklist work of the 2000s. Industrial controls have been doing it since the international industrial-security standards started getting taken seriously in the early 2010s. The list is longer than this chapter has room for.

There is a particular wrinkle in the cybersecurity version that is worth pulling out, because it changes how to read where AI governance is heading.

Privacy and cybersecurity took different regulatory paths over the past three decades. Privacy got laws. GDPR in Europe, CCPA in California, the patchwork of state laws that followed. The laws defined consent, data minimization, and rights of access as enforceable obligations on the firm. The cost of building privacy in became lower than the cost of bolting it on, because the law made bolting it on (or ignoring it altogether) expensive.

Cybersecurity got the opposite. No broad design-time mandate. Practitioner discourse, vendor tooling, voluntary frameworks, breach-disclosure laws that work as failure-mode regulation rather than design-time mandate. Result: privacy is now built-in at the data layer in most large enterprises but cybersecurity, after thirty years of secure-by-design literature, is still bolted on at the end of many IT projects. If anything, the pattern is sharpening. Vibe-coded applications, built fast and shipped faster, very frequently have security bolted on after deployment, when something goes wrong (if then).

So far, AI governance is reproducing the cybersecurity pattern. Practitioner-led discourse. Vendor-supplied tooling. Voluntary frameworks. Breach-style regulation that operates after a failure. No broad mandate that built-in be the default. This matters for the executive choice in front of you. The regulatory environment is not going to require built-in. The default it is creating is bolt-on (just as with cybersecurity). If your firm ends up built-in, it will be because you decided that. If your firm ends up bolt-on, it will be because the default carried you there.

Two cases from the human-paced era anchor what a mismatch looks like when the gap gets allowed to run.

Wells Fargo's cross-sell sales program, branded "Eight is Great" inside the firm in the years leading up to the scandal, worked acceptably for years. The audit machinery was designed to look for outliers in branch-level sales behavior. It worked because most branch employees were not opening fraudulent accounts. Through the early 2010s, the sales pressure escalated. Incentives sharpened. Branch managers were measured, and managed, against cross-sell targets that had become difficult to hit honestly or ethically.

By the time the Office of the Comptroller of the Currency issued its consent order in September 2016, the wrong behavior was no longer the outlier. The audit was built to catch a few dishonest employees in a workforce of mostly honest ones. By then the dishonesty had spread across the branch network, pushed by the incentives. A system built to spot rare exceptions could not see a problem that had become the rule.

Boeing's 737 MAX is the second analog. It was a new airplane certified as an update to an old one. Pilots moving up to it were trained on the systems earlier 737s had, on the assumption that they already knew how the airplane behaved. Boeing then added a new automated system, the Maneuvering Characteristics Augmentation System, or MCAS, which could push the airplane's nose down on its own. For reasons that were partly about avoiding the cost of retraining pilots and partly bureaucratic, MCAS was left out of what pilots were told to expect. They were flying an airplane that could do something they had not been trained to work with.

The built-in safety culture that Boeing had relied on for decades, the one that worked because pilots knew the automated systems they were flying with, could not see what the team's training and incentives (its overall operating context) were producing. The September 2020 House Transportation Committee report on the two MAX crashes and the February 2021 Inspector General audit converged on the same finding. The boat had been redesigned. The chart guiding it had not been updated. The crew was navigating against the old chart.

Worth pausing on what kind of failure each of these is. Wells Fargo is the lookout scanning for the wrong thing: bolt-on failing. Boeing is the chart never redrawn while the boat changed underneath it: built-in failing. And Boeing's chart had been a good one for decades, which is exactly the trap: built-in is not something you install once. The chart only protects the boat it was drawn for, and as the boat keeps changing, the chart has to keep being redrawn. A stale chart fails the same way a misaimed lookout does, only more quietly, because everyone aboard believes the protection is built in.

The Wells Fargo gap ran for years. The Boeing gap ran for years. A governance failure at human pace produces, perhaps, a few mistakes a week, sometimes for a long time, before anyone catches the pattern. The same failure mode at agent pace can produce a thousand mistakes in the same span. The Air Canada, Microsoft Copilot, and Deloitte cases earlier in this chapter are sketches of the agent-paced version while it is still small. The open question is what the same pattern looks like at scale, after twelve more months of agentic deployment compress what used to be human-pace decision throughput into machine pace.

The river was always there. The current is much faster now.

Direct about what's mixed

The case for bolt-on failures is thicker in the public record than the case for built-in successes. Three reasons. Agentic deployments are new, and built-in architectures have not had time to generate the kind of public track record that older arrangements have. Firms do not publish the details of their internal governance arrangements unless their market positioning rewards it, which is why the most polished built-in narratives come from professional-services firms and a bank that have made platform positioning part of their market story. And successes do not generate tribunal decisions, civil-resolution claims, or vulnerability disclosures.

The asymmetry is in the data. It does not reflect the underlying frequency. On the built-in side of this chapter, the firms named are firms with deployments that lean built-in for the work being described. The firms themselves run a mix.

One caveat before moving on. The bolt-on versus built-in distinction is a first cut, and a deliberately coarse one. Researchers who work on AI governance full-time will tell you the binary is too coarse, that the real variables are things like who holds authority, how much autonomy is granted, and where accountability lands. They are right. Those finer distinctions matter a great deal when you are building the system. They matter less when you are the executive deciding which way to lean for a given piece of work. That decision is the one this chapter is built to help you make, and for that decision, two clear options beat five blurry ones.

• • •

Key takeaways

- **Supervision, autonomy, and governance are three views on a single question.** How does work get done well when it could go wrong?
- **Bolt-on catches problems after they occur. Built-in keeps them from occurring.** The right approach is the one that fits the work.
- **Built-in governance is highly related to context provision.** Just viewed from a slightly different angle.
- **Most firms run both approaches across different deployments.** Set the approach deployment by deployment.
- **A governance gap at human pace produces a few**

mistakes a week. At agent pace, the same failure can produce a thousand. The choice is old. The speed is new.

• • •

The boat now has its design, its context, and its governance, deployment by deployment.

What you do not yet have is the discipline for running the three Tensions together as the water keeps moving. The right setting for any one of them does not hold still, because the river does not hold still.

The work of getting the three right is not a project that ends. It is a practice you keep at.

That practice is what comes next.

Further reading

Companion website: riverdoesntwait.com

A few readings to go deeper on this chapter. Sources covered in earlier chapters are not relisted here.

Moffatt v. Air Canada, 2024 BCCRT 149

The British Columbia Civil Resolution Tribunal decision quoted in the opening of this chapter. Worth reading in full as the fullest available primary-source articulation of what happens when an enterprise treats an agent as legally separable from the rest of the operation. The decision is short, the language is direct, and the finding (that an agent is part of the firm's operational surface area whether the firm has set it up that way or not) maps to the chapter's argument with no translation required.

Mindful AI Foundation publications

Four pieces from the Mindful AI Foundation develop the architectural case for built-in governance in AI systems specifically. The peer-reviewed entry point is Mikkilineni and Kelly (2025), "From Static Prediction to Mindful Machines: A Paradigm Shift in Distributed AI Systems" (Computers 14(12), 541, MDPI). The paper is the source of the "coherence debt" term used in this chapter. Three additional pieces (a manuscript on governance of commitments, and two book-in-progress extracts) extend the argument in different directions and are available through the Mindful AI Foundation. Of the four, the Foundation's case is the maximalist version of the built-in argument and is the right reading for executives who want the architectural case at full length.

Derek Waldron interview with McKinsey, "JPMorgan Chase's Derek Waldron on building an AI-first bank culture" (October 2025)

The prompt-engineer-to-context-engineer framing in Waldron's own voice. The full interview covers the emergence of new job categories at JPMorgan and the firm's knowledge-management approach to making institutional information available to AI. The chapter borrows a single sentence; the rest of the interview is worth reading in full.

Matt Wood on the PwC One launch (Wood LinkedIn post, March 19, 2026; PwC corporate press release)

The most authoritative single articulation in the verified executive-voice corpus of the bolt-on / built-in distinction. Wood has since moved from PwC to AWS, where he is Chief AI and Technology Officer; the launch-day framing is read with role-at-the-time attribution. The PwC press release carries the institutional articulation; Wood's LinkedIn carries the personal voice.

Simon Willison, "The lethal trifecta for AI agents" (simon willison.net, June 16, 2025)

Willison coined the term to describe the architectural conditions that make prompt-injection attacks like the Microsoft 365 Copilot one in this chapter possible. Three properties in combination produce the danger: access to private data, exposure to untrusted content, and the ability to externally communicate. The framing is the plainest vocabulary in the current public record for the configuration that built-in architectures address by cutting one leg of the trifecta rather than monitoring for the combination.

Bain & Company, "The Three Layers of an Agentic AI Platform" (April 2026)

The closest pre-existing articulation of the specific bolt-on / built-in distinction in the consulting-firm literature, with the "embedded by design, not bolted on after deployment" framing applied to AI specifically. Useful as evidence that the distinction is in circulation among advisors to the executive audience this book is written for.

Chapter Seven

The Plan You Keep Remaking

Why the strongest position you can take is one you fully intend to change

Sooner or later, someone is going to ask you the question directly. It might be a board member, three items into an otherwise routine meeting. It might be your CEO in a hallway. It might arrive as a slide your strategy team built without you, titled "Our AI Roadmap," which you are now expected to defend. However it comes, the question is the same. What are you actually doing about all this?

You have read six chapters. You know the capability curve is real, that agentic proficiency keeps climbing and the leaders are already pulling away from everyone else. You know the choice between retrofit and reimagine is one your organization is making whether you make it on purpose or not. You know the river is moving and the right position on it is something you navigate continuously. You know the three Tensions you have to get right inside the boat relative to the goals you need to achieve. What you may not have yet is an answer to the question in the room, in language the room will accept.

The instinct, at this point, is to do the thing you have done your whole career. Devise a strategy. Make a plan. Commit to it. Execute against it. Reassess it next year, or next quarter if the situation feels urgent. That instinct is the one thing this chapter is going to take away from you, because it is an instinct that no longer works.

Start from what you are actually responsible for

Before we get to what to do, one correction to the frame, because it changes everything that follows.

You are not managing AI. You do not have an AI problem, an AI strategy, or an AI roadmap, whatever the slide says. You are responsible for a set of business outcomes. Cycle time in your operation. Cost to serve. Customer retention. Risk reduction. Profitability and growth. Whatever sits on the scorecard you are measured against. Those outcomes have not changed because of AI. What has changed is that the way you achieve them is now in constant motion, because the medium you achieve them through has started moving under you, faster than ever before.

That reframe is not cosmetic. It decides which of two places you end up. Treat this as an AI initiative and you get an AI initiative: a task force, a center of excellence, a portfolio of pilots that generate impressive activity and move no number anyone cares about. But keep your eye on the outcome and agentic AI becomes what it actually is, a faster and more capable way to reach the results you are responsible for. The initiative produces motion. The outcome moves the number you are measured on. This chapter is written for the second of those, the outcome, and the first step onto it is to stop asking what your AI strategy is and start asking what the current that AI is accelerating is doing to the outcomes you already own.

The three decisions you already know how to make

If you read the three Tension chapters over the course of a few weeks, the first of them may already be a blur. So I'll put them back in front of you quickly, because you are about to use all three at once. In practice, you work all three at the same time, on the same piece of work.

The first Tension was workflow and workforce. The question was whether you slot agents into the org you already have, into the existing seats and handoffs, or whether you redesign the work and the roles inside it together. Agents in old seats is the slow path to irrelevance, because it captures a sliver of the available value and leaves the structure that limits the value untouched. Reshaping the boat and reconfiguring who is in it is a single decision. That was Chapter 4.

The second Tension was context. As the work is being redesigned, the people and agents doing it will likely need to see things the old work never required them to see. How much of what you know, what the firm cares about this quarter, what the real constraints are, do you make explicit so the team can act on it? The reality is that the scope of context you can provide turns out to be the scope of organizational influence you actually hold, which makes context provision a more personal activity than it first looks. That was Chapter 5.

The third Tension was supervision, autonomy, and governance taken together. How much does the team do without a human checking each step, and what keeps a wrong move from becoming a wrong outcome? The choice was between catching problems after they happen and shaping the context so the problems mostly do not happen in the first place. Bolt-on governance for work that still looks like the old work. Built-in governance for work you have genuinely redesigned. Match them to the wrong kind of work and you fail in predictable ways. That was Chapter 6.

Those three Tensions are not three separate decisions. They all come down to a single judgment, the one this whole book has been about. **How far toward reimagine does this particular piece of work need to go?** Answer that, and the three Tensions stop being three open questions and become three distinct, highly-linked settings that follow from the answer. Work you are retrofitting wants agents in adapted seats, context bounded to the task, governance bolted on as a check. Work you are reimagining wants the roles redrawn, context broad enough to cover the new seams, governance built into the substrate. The retrofit-or-reimagine judgment is the steering. The three Tension settings follow from it. Follow, not in sequence. The judgment sets the position; the three settings get tuned together to match it, each along its own spectrum.

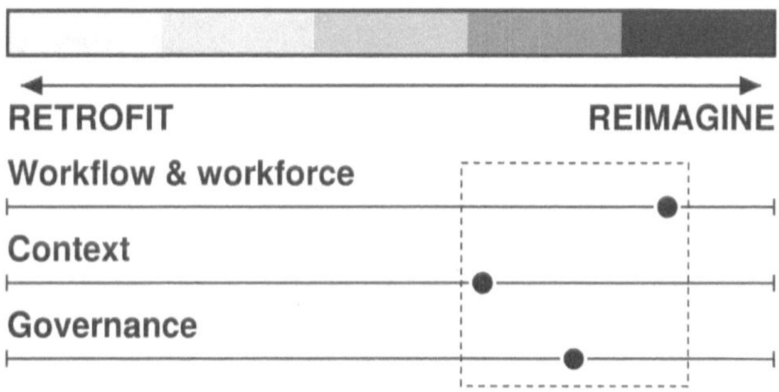

Illustrative example of how the three Tension settings are distinct yet strategically aligned with the desired position on the retrofit-to-reimagine spectrum

So the operational question is never "how do I handle the governance Tension." It is "how far toward reimagine does this work need to be, and therefore where do the three Tension settings need to sit." Get the steering judgment right and the Tension settings follow. Get it wrong, position as though you are retrofitting work the current has already pushed into reimagine territory, and you get exactly the failures Chapter 6 catalogued: the lookout scanning the horizon while the rocks sit under the bow.

Not all your outcomes are in the same water

So far this sounds like something you do to one piece of work. In reality you own several outcomes at once, and they are not all in the same water.

A river runs slowly through a wide flat stretch and fast through a narrow gorge. The current of agentic capability does the same thing across your portfolio of outcomes. Some of your outcomes sit in fast water, where the technology has already pushed the work hard toward reimagine and competitors who have redesigned are pulling visibly ahead. Document-heavy work. Coordination across many systems. High-volume exception handling. Knowledge that is fragmented across the organization and expensive to assemble. If one of your outcomes depends on work like that, the current under it is strong, and the cost of holding your old position is rising every quarter.

Other outcomes sit in slower water, where error costs, human trust, or the texture of a human relationship dominate the work. High-stakes judgment. Clinical or advisory encounters where being wrong is catastrophic. Work where the conversation itself is the product. The current is real there too, but the flow is gentler, and moving prematurely costs more than waiting another year for the technology and the controls to mature. (Having said that, these kinds of outcomes are also where some of the most interesting reimagination will eventually become possible.)

This is the triage, and it is the part you can take into the strategy meeting or boardroom. You are not going to reimagine everything at once, and you should not pretend to. You identify which of the outcomes you own are in fast water, steer those hard now, and let the ones in slow water ride on a lighter setting while you keep watching them. A board can hear that. A board can do nothing useful with "we are transforming with AI." It can fund and hold you to "these two outcomes are in fast water, here is how far I am taking them this year, and these others I am watching but not yet moving against."

The discipline is to make the call on purpose. Most organizations do not. They move on the outcomes that are easiest to move, or most impressive to demo, while the water moves fastest somewhere they are not looking. Reading the water first, and steering where the current is genuinely strong, is most of what separates the executives who are pulling ahead from the ones generating activity.

Your first moves

Out of that reading comes your first concrete moves. They come in two kinds.

The first kind is the position you commit to. For each outcome you have decided to navigate towards, you are making a call about how far toward reimagine it goes this year, and therefore where the three Tensions sit: what work gets redesigned and who staffs it, how much context the team gets and at what scope, how much autonomy the team runs with and what catches a bad call. That is a position you can state in a sentence per outcome, and it is exactly the thing the room has been asking you for. Not "our AI strategy." Instead: here is the outcome, here is how far I am taking it, here is what that requires.

The second kind of move is the help you ask for, and this is the one executives skip at their peril. Reimagining how an outcome gets achieved almost never sits inside one person's authority. Redesigning a workflow that crosses three functions needs the executives who own the other two in the room. Giving a team the context it needs to act often means reaching for information, systems, or relationships that belong to a peer. Building governance into the substrate is work you do with your CIO or CTO, not at them.

So part of your first move is naming, specifically, where you need help and from whom. From your CEO or Chairman: air cover, and the mandate to redesign rather than optimize. From your CIO or CTO: the platform and the built-in controls that let a team run wider without running reckless. From your peer executives: the parts of the context that cross your boundary into theirs.

This is where the earlier point about convening your peers becomes a literal action item. The executive who treats this as a solo exercise, redesigning only the part of the outcome they fully control, produces a redesign with seams exactly where their authority ends. The ones who pull their peers into the same conversation, ideally working from the same framework, get a redesign that holds across the boundaries. If there is a single highest-value thing to do the week after you finish this book, it is probably to get the two or three executives who share your most important fast-water outcome into a room and read the water together.

The plan you intend to keep remaking

Here is where the career instinct comes back to bite, so I'll name it and retire it.

Everything I just described, the read, the position, the first moves, looks like a plan. You could write it down, take it to the board, and feel that you have done the thing. And you have, for now. The trap is this: the plan is real, but the plan is provisional by design. Not provisional because you did it badly. Provisional because the water keeps moving, and a position that was right this quarter will be a little wrong next quarter and meaningfully wrong by the time your annual planning cycle comes around again.

This is the one habit worth breaking on purpose. Call it the snapshot mindset: the belief that you deploy, then reassess on a schedule, the way you commission a building and inspect it once a year. It is the most natural thing in the world for an executive trained to plan a year out and defend the plan, and it is the one habit that risks failing hardest in the context (pun intended) we have been discussing. It fails for the same reason agents-in-old-seats failed in Chapter 4 and bolt-on governance for reimagined work failed in Chapter 6. It freezes something that is supposed to keep moving. A snapshot of a river tells you nothing about where the river is going or how it behaves.

There is a sharper way to see it, one that sits right on the three Tensions. When you make the retrofit-or-reimagine call for a piece of work, you adjust the three Tension settings to match the water as it is. The snapshot mindset treats those settings as an anchor, dropped once and assumed to hold. But the river keeps moving, and a boat anchored to last year's settings does not react the way you expect. It gets buffeted by the current, and the longer the anchor holds, the more likely the anchor will break. And that never leads to good outcomes.

What replaces it is not faster planning. As a startup executive who has worked alongside large enterprises for three decades, I learned the hard way that a Global 1000 company cannot rebuild its operations every quarter, and I am not going to pretend otherwise. The replacement is to split one activity you have mostly treated as a single annual event into two activities that run at two different rhythms.

The first activity is watching. Sensing whether your current position still fits the state of the river. Watching is cheap. It costs attention, and little else, and it can and must happen often, because the river does not wait for your planning calendar.

The second activity is the move you make when watching turns up a real change. It is heavy and deliberate, and it takes one of two forms: adjusting the three Tension settings, or re-aiming the boat at a new destination. Either way it is expensive and can be disruptive, it usually has to happen in the middle of operations, and a large enterprise can only absorb so much of it at once. It happens on a slower rhythm, by necessity.

The one rule that matters is the relationship between the two. The watching has to outpace the moves it triggers. You sense at the speed of the river and you commit at the speed your organization can actually absorb. Put plainly: you watch all the time, and you make the big moves only as fast as the organization can take them. I am deliberately not providing you a rule such as "the watching is weekly and the moves are quarterly," because the right cadence depends on your organization's metabolism, which you know and I do not. What I am telling you is the ordering. If your sensing runs at the same slow rhythm as the moves it triggers, you are back in the snapshot, just with extra steps.

And here is the part that should be a relief rather than a burden. The failure mode of large enterprises in this environment is not that they change too slowly. Of course they change slowly. They are large - sometimes supertankers. **The failure mode is that they sense too slowly**, so by the time the annual cycle comes around they are reacting to a river that moved two channels over while they were not looking. Sensing faster is the agility move that is actually available to you. It is cheap, it does not require you to become a startup, and it is the thing that makes your slower, heavier moves land in the right place instead of the wrong one.

The questions you keep asking

So what does the watching actually consist of, in practice? Not a checklist. A checklist is a snapshot wearing work clothes: a list of tasks you complete and then file. The thing you want is a small set of questions you keep asking, the same questions, on a regular rhythm, against each outcome you are steering. The questions do not get completed. They get re-answered, and the answer is allowed (encouraged) to change.

There are only three things that can make your position on an outcome wrong, which means there are only three things worth watching for. The technology can change. The competition can change. And your own earlier read can turn out to be wrong. Three questions, one for each, and each one light enough to ask often.

The first question is about the technology. **Has a capability crossed from demo to deployable against the actual work behind one of my outcomes?** Stated for a team that has not read this book: is there a new tool or technology that can really do this work now, not just look good in a demonstration? The distinction matters, because most of what gets announced is not yet usable against your specific work, and acting on a demo is how you end up with a pilot that impresses everyone and changes nothing. Recall from Chapter 3 that a surge in the river is not an AI model release. It is a capability meeting a way to deploy it against work you can structure, with controls you can defend, all arriving in the same window. That convergence is what you are watching for. The press release is not the signal.

The second question is about everyone else. **Has someone reset what good looks like for one of my outcomes?** For the team: has a competitor or a peer changed the bar on this? Picture a manufacturer whose biggest competitor brings up a fully agentic-AI-run, lights-out plant that operates several times faster at a fraction of the cost. That is not an AI development to the executive who runs manufacturing. It is a manufacturing-efficiency development that happens to be AI-enabled, and it matters for the same reason any competitor doubling its cost advantage matters. You are not watching for AI news. You are watching for the moment someone makes your current level of performance look slow, and noticing that those moments increasingly arrive in AI-enabled form, faster and less predictably than they used to.

The third question is about you. **Has something I judged could wait started to move faster than I expected?** For the team: is anything we decided not to worry about yet actually speeding up? Every triage you make is a bet that certain outcomes can wait. The current does not care about your bets. The work you confidently set aside last year, because the technology was not ready or the error costs were too high, is exactly the work most likely to surprise you, precisely because you stopped looking at it. This question is the discipline of auditing your own earlier judgment before the market audits it for you.

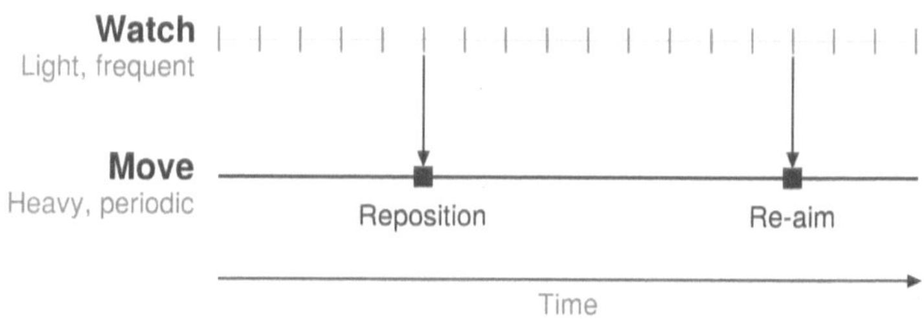

Continuous strategic navigation with two rhythms

Keep the watch pointed at your outcomes

Notice what all three questions have in common. Each one points at outcomes you own. Keeping the watch pointed there is the most important discipline in the whole activity, because without it the watch drifts into exactly the thing you do not need: a running scan of all of AI rather than the parts of it that touches your most important outcomes.

The easiest way to feel responsible while accomplishing nothing is to form a standing committee that reports, every thirty days, on every new model, every new agent framework, every benchmark that moved. Maybe an AI Newsletter that gets sent to the executive team? That is not navigation. That is a weather report for water you are not on. A new model release is a signal to you only if it changes how far toward reimagine one of your outcomes now needs to go, or could go. If it does not touch an outcome you own, it is noise, however impressive the demonstration. The outcome is the filter, and the filter is what turns an infinite scanning problem into one that a human executive can actually carry. You are not becoming an AI-watcher. You are watching your outcomes, and the radar you already run simply needs to reach further and refresh more often, because the things that move your outcomes are increasingly AI-shaped.

There is also the question of who does the watching, and it has only one right answer. The watch cannot be handed off to a market-intelligence team, a strategy consultancy, or an industry-analyst subscription (no disrespect implied to my former home, Gartner), or a research agent you set up. Those can feed it, and they are genuinely useful for that, but they cannot be it. An outside party can tell you that a capability now exists. Only the people steering toward an outcome can tell whether it has changed the water around that outcome, because that call depends on knowing the work from the inside. Sensing that gets delegated out of the operating seat quietly turns back into a report you read on a schedule, which is the snapshot you were trying to escape. The watch belongs to the people navigating the boat.

One deliberate exception keeps the outcome filter from blinding you. Most of your watching is tight to the outcomes you are already steering. But every so often the current opens a channel you were not pursuing, and makes reachable an outcome that was not on your list. The lights-out plant does not only threaten your cost position. It raises the question of what your manufacturing operation could now become. That lateral surprise is rare, and you should not let chasing it pull you back into scanning everything. But leave a small, deliberate slot in the watching for whether the river just put a new destination in reach. That question is where reimagining the outcome itself begins, and it is the larger of the two ways you respond when the water changes. The two ways are worth taking one at a time.

When the water changes, adjust the settings

Watching is how you notice. The move is how you respond. When the watch shows your position on one of your tension settings is no longer right, whether the technology moved, a competitor moved, or your own read was wrong, you have a move to make, and most of the time it is not to throw everything out. It is to adjust the settings you already chose. Stated plainly for the team: this changed, so we need to change how we have this set up.

And "how we have this set up" is precisely the three Tensions, which is why everything you worked through in the last three chapters comes back to do its job here. Each of the three Tensions exists along a spectrum running from retrofit to reimagine, and the position you took on each one was a choice you made for the water as it was. Adjusting the settings means moving each one along its spectrum to match where the water is now. It is not three new decisions. It is revisiting the three settings your steering judgment produced the first time, now that the water has moved. The order below is an order of asking, not an order of doing. You will still commit the three settings together.

Start with workflow and workforce. If the water rose, the question is whether the redesign you did last time went far enough. Work you slotted simple agents into a year ago, sensibly, because that was the right call for where the current was then, may now be work the current has pushed toward genuine redesign. Adjusting the workflow-and-workforce setting means asking whether more of the work should be restructured and more of the roles redrawn than you were willing to commit to before. For the team: do we need to change who does this work and how it is organized, instead of attaching a better tool to the way we already do it?

Then context. As the work gets redesigned further, it crosses more boundaries, and the people and agents doing it need to see more than they did. Adjusting the context setting means widening what you make explicit, and often reaching into territory a peer controls, which is exactly why this is the moment to convene the peers who share the outcome rather than to quietly expand on your own. And the widening does not always stop at your own walls: the context a redesigned piece of work needs can reach past the organization entirely, into partners, suppliers, or outside signals, the widest of the three levels of context from Chapter 5. For the team and those peers: as this work changes shape, what do the people and agents doing it now need to see that they could not see before, and who owns it, inside the company or beyond it?

Then autonomy and governance. As you give the redesigned work more room to run without a human checking each step, the controls have to move with it, from catching problems after the fact toward keeping them from happening at all. Adjusting the governance setting means moving a piece of work further from bolt-on checks toward built-in controls as its autonomy widens. For the team: as this work does more on its own, what keeps a wrong move from becoming a wrong outcome, and is that protection built into how the work runs or just inspected afterward?

That is the whole of the adjustment: the same three settings, nudged further toward reimagine because the current moved. It is not a transformation program. It is the adjustment a navigator makes when the water shifts: resetting the boat for the new current rather than running it on settings the river has outgrown. It is the move you will make far more often than the larger one that comes next.

When the destination moves

Once in a while the change is bigger than an adjustment. The current does not merely push your work further toward reimagine. It puts a different destination within reach, and the right move is to re-aim the boat: to change what you are aiming at. This is the deepest version of the choice this book is about, and the rarest.

Most of the time you are reimagining the path to an outcome that itself stays fixed. JPMorgan did not decide to stop running know-your-customer checks. The firm decided to do the same regulated thing in a radically redesigned way, and took the cycle from days to under a minute. The outcome held. The path was reimagined. That is the common and powerful case, and most of your reimagining will be this kind.

But sometimes the technology makes a genuinely different outcome reachable. Block did not reimagine how to run a management hierarchy more efficiently. The company reimagined what it was organizing around, from a hierarchy to something it describes as an intelligence. Moderna did not make its HR function and its IT function each a little better. It collapsed them into a single function organized around architecting the flow of work, an outcome that did not exist on anyone's org chart a few years earlier. Those are reimagined outcomes, not reimagined paths.

This is the higher-altitude move, and it is the one to reserve for the slower rhythm, because re-aiming is not something you do while steering hard toward the destination you already chose. Day to day, the outcome holds still and you navigate toward it. Periodically, you lift your eyes off the water and ask whether the shore you aimed at is still the best shore available. Most of the time the answer is yes. When it is no, that is the most consequential move you will ever authorize.

Note: Reading the Water Together, at the back of this book, provides a discussion guide that you and your team can use to further operationalize the concepts discussed here.

• • •

Key takeaways

- **You are not managing AI.** You are achieving the outcomes you already own, through a medium that now moves under you, more quickly than ever before.

- **The retrofit-or-reimagine judgment is the steering.** The three Tension settings follow from it, not three separate things to navigate.

- **Not all your outcomes are in the same water.** Steer

hard where the current is strong, and let the slow water ride while you keep watching it.

- **Retire the snapshot mindset.** The plan is real and provisional by design; a snapshot of a river tells you nothing about where it is going.

- **Sense at the speed of the river, commit at the speed your organization can absorb.** Watch the technology, the competition, and your own earlier read, and move as fast as the organization can take it.

- **The move takes two forms: adjust the settings or re-aim.** Adjusting the settings nudges the three Tensions to match the new water; re-aiming changes what you are aiming at, and is the rarer and more consequential move.

- **Convene your peers.** The highest-value first action is getting the two or three executives who share your most important fast-water outcome into a room to read the water together.

• • •

The real answer to the question in the room, the one a board member or your CEO is going to ask, is not a roadmap, and it is not a reassurance that everything is under control. It is something steadier than either. You know which of your outcomes are in fast water. You know how far toward reimagine you are taking each one this year, and what that asks of your workflow, your context, and your governance. You know where you need help and from whom. And you know that you will be back, because you intend to keep watching the water and keep remaking the plan as it moves. That is a thing you can say out loud, and mean, and act on Monday morning.

That is the core of it. If you closed the book here, you would hopefully have what you came for: a way to think about the change your organization is moving through, and a way to act on it that does not pretend the change will hold still long enough for a tidy plan.

There is one more chapter, and it is a different kind of chapter. The whole of this book, up to now, has been about your organization. Your outcomes, your boat, your river. The last chapter is about you. Not the executive. The person. What all of this does to a career, to a sense of what your work is for, to how you lead when half the team is no longer human. It is the most personal thing in the book, and you can take it or leave it. The work here is done. What follows is the personal part.

Further reading

Companion website: riverdoesntwait.com

No new sources here. The goal in this chapter was to put to work the cases and arguments built up across Chapters 1 through 6: the capability trajectory and the leader-laggard gap, the retrofit-or-reimagine choice, the river and its surges, and the three Tensions with their named cases (Block, Moderna, and JPMorgan on the reimagine side; Air Canada, Microsoft 365 Copilot, and Deloitte on the governance-failure side). All of them are in the earlier chapters' Further Reading sections, and the companion website aggregates everything in one place. If you want to go deeper on anything here, go back to the chapter where it first appeared.

In the Quiet

Epilogue

You could have closed the book a chapter ago.

By the last page before this one, you had what you came for. You have a framework for knowing which of your outcomes are in fast water and which can ride the slow current a while longer. You know how far you mean to take each one this year, and what that asks of your workflow, your people, and your controls. You have some new ideas for how to lead your organization through this wave. That was the deal I offered you at the start. An operator's read instead of a vendor pitch.

So what comes next sits outside that deal. I am taking the advisor's hat off for a few pages. What follows is a different kind of writing, closer to the things I find myself turning over when the calls are done and the dashboards are closed and it is just me and the scale of what is happening around us. The conclusions are yours to draw. I am only going to tell you what I tell myself, because in the ways that matter, I am probably not too different from you. Late fifties. A working life spent building teams and carrying the weight for the people on them. Watching waves of technology arrive, and working out what they mean. Sometimes even helping to create the wave.

For me, this time. The person underneath the title.

And the first thing surprised me.

For most of us the fear of all this change runs deeper than money. We will be fine. If you picked up this book you are likely well into a good career, you have provided for others, and your retirement account is safe from any agent. The real fear is quieter, and harder to say in a room full of peers. It is that the work might start happening without us. That the particular thing we got good at over twenty or thirty or forty years, the judgment, the read of the room, the call nobody else wanted to own, is turning into something a machine now does faster and more competently than we ever did. And if the work was, somewhere along the way, a large part of who we became, then the real question underneath all the strategy turns personal. It is this. Who am I, when the thing I was can be done by something other than me?

I feel that one, and peace with it is still a way off. I have spent a career being the person who could hold the whole picture in his head at once, and I now work every day with systems that hold more of the picture than I ever could. That is an uncomfortable sentence to write.

So I'll be plain about what lives through this and what falls away.

Judgment is moving. Strategy is moving. The parts of this job I was proudest of, the synthesis, the pattern I could see a beat (or a week) before the room did, are precisely the parts that a system carrying a thousand times my working memory can now do, and will keep getting better at. Pretending those parts stay ahead of the machines is the one lie I have refused to tell you anywhere in this book, and I will keep refusing it here, on the last pages, even to make us feel better.

What is left is smaller than I would like, and more important than I used to believe.

Someone has to be accountable, and accountability lands on a person. When it goes wrong, and it will, someone has to sit in front of the board, the regulator, the person who got hurt, and answer for it. That someone is you. That answerability sits outside anything you can automate, because it was never really a task. It is most of the reason the seat exists. It was always human. It stays human, because it is made of being a person who can be held to account.

Trust holds too. The real kind that runs between people, a different thing from the words on the slide. What lets your leadership team take a hard turn together and stay whole is the years you have spent learning whether you can rely on one another, and that lives in people, beyond the reach of any system. The reason a customer or a partner stays through a bad stretch is that they are dealing with a person whose word has weight. A person's word stays a person's, and I expect it always will, because the whole value of it is that a person staked it.

And people still follow people. Through the kind of change you are about to lead, what carries an organization is one thing above the elegance of the plan: whether the people being asked to change believe the person doing the asking. That is leadership, and it is the making of meaning, and it is as human as it has ever been. More so now, probably, because there is more change coming at us than at any point in our careers.

There is something underneath that I feel most and have rarely verbalized to others before now. By this stage, for a lot of us, leading people is not just the hardest part of the job. It is the part I love most. Somewhere along the way the number stopped being the thing that got me out of bed, and the people became it. The rarest moment in a working life, rarer than any record quarter, is the one where a team turns to you at the end of something hard and says, and means, thank you for taking us there. We did not think we could. I have had that a handful of times in thirty years, and nothing the work ever gave me comes close.

So here is the question I have not finished answering for myself. How ready am I to get it less often, or to get a thinner version of it from a machine that did the task beautifully and felt nothing at all, that finished the job and never looked up?

Because the strange part is that the skill carries over but the reward does not. The thing I had to get good at to rise, setting a direction and giving people what they need to move toward it on their own, is now also how you get the best out of the machines. It is needed more than ever, not less. But the satisfaction was only ever on one side of it. It came from the people. An agent will take your direction and outperform on it and give you nothing real back, because there is no one in there to do the giving.

And that tells me the answer. Tells me where to spend myself. If the meaning was always in the people, then carrying them through the most disorienting change of our working lives is not the overhead around the real job. It is the job. They are more uncertain, and some of them more afraid, than at any point in the years we have been leading them, which means they need someone worth believing more than they ever have. It may be the largest chance I will ever get at the one thing I am proudest of. Take a team through this, to a place none of you believed you could reach, and the thank you on the far side of it could be the one you remember longest.

Which is the thing I actually wanted to say, and the reason I kept writing past the end of the last chapter.

We are in the habit of treating legacy as something separate from the work itself, a small set of outcomes we recite at the send-off, the record quarter, the integration delivered, the deal closed. I doubt that now, and I wonder if I ever truly believed it. Your legacy is the work itself and its human impact, seen from far enough back that the shape of it shows. And the shape is as much how you got there as what you reached. The team you built up or hollowed out on the way to the number. Whether you would do it the same way again. It was never a separate outcome waiting at the end.

Every decision you are now going to make about this technology is a small moral choice in operational clothing. Which work you give to a machine and which you keep in human hands. Who you stand behind when a role dissolves, and who you let slip out quietly. Whether the people coming up behind you inherit an organization that still has a ladder worth climbing, or one you optimized so cleanly on your way out that the ladder is gone. The framework you just absorbed is silent on every one of those choices. And all of them are the framework, lived out across a career.

I think about the people I hired most recently who are approaching thirty now. They are going to spend their entire working lives inside the world we are building in these next few years, in the decisions we make half-distracted, between meetings, under pressure to show a number. We are the last people in the building who will remember how it used to work. That is closer to a privilege than a weight. For a short while longer, we are the ones who get to decide what gets carried forward and what gets left behind.

Here is the thing I catch myself doing, which is why I have learned to be a little wary of myself.

This whole book has been about getting good at the navigating. Reading the water, moving when it moves, taking the boat further than you dared to last time. And being good at a thing has a quiet way of becoming the reason you do it. I notice the pull in myself, to automate something simply because I can, to reach for the cleaner version of every process, to treat the human friction in how work actually gets done as a defect to be engineered out. I do it for the pleasure of it, freely, because navigating well is satisfying, and the satisfaction starts, if you let it, to feel like the point.

And it was never the point. The machines rising up in some scene from a film were always the wrong thing to fear. The risk I actually worry about is duller and more plausible. That we will get so good at the steering that we forget to ask where we meant to go. That we will hollow out the human parts of our organizations because we can, and because the spreadsheet rewards it, and call the hollowing progress, and notice what we gave up only once it is gone for good.

The technology is indifferent to all of it. The credit, or the fault, will be ours. The pro-human choice is a vote for something. It is a choice you make inside the building, one deployment at a time, with the room empty and your hand free. That is the uncomfortable part. The decisions that matter most arrive in disguise, looking like default settings, and you will have to choose to look at them.

And there is a larger version of this, too. In the quiet, I believe it runs far past the edge of any company.

I have worked to keep this book inside the walls of the organization you run, because that is where you have real leverage and where I have something useful to offer. But the truth, on the last pages, is that the questions run well past those walls. Every one of these calls, which work stays human, who gets protected, what we are willing to hollow out for the sake of efficiency, is being made right now at a far larger scale. By governments. By whole societies. By a species, all at once, made by people just as fallible, and just as hurried, as you and me, standing in the same fast water and reaching for the same defaults.

My enterprise is a microcosm. So is yours. We are each running, at a scale small enough to hold in two hands, the very experiment the world is running at a scale that outruns any single pair of hands. That is the biggest thing that keeps me up at night. My own eventual 'retirement' dropped off the list a long time ago.

So I come back, the way you have to, to the one thing I can actually put my hands on. The world was never mine to steer. The part of it I was given is. And that part turns out to be larger than it looks, because it is a working model of the very thing that runs past my steering, and how I handle it is the only vote I am sure I get.

That is what I tell myself when it is quiet. Get good at the navigating, because the navigating is real and the people who depend on you are counting on you to be good at it. But hold it in its place. The reason was always the people. It was always about what we hand to the ones who come after us, and whether they will be glad we were the ones steering when the water turned.

The work here is (finally) done. The rest of it is yours.

Reading the Water Together

A Discussion and Reflection Guide for The River Doesn't Wait

A note on using this guide

This guide is built for a room of senior leaders, each of whom owns a boat of some size and is trying to navigate it down the same river. It is not a quiz on the book, and it is not designed to land everyone in the same place. The book refuses to tell you the right answer on purpose, because there isn't one that holds still. A downloadable version of this guide can be found at the book's website: https://riverdoesntwait.com/

A few things to set up before you start:

- **Two kinds of questions.** *Reflect* prompts are for thinking about on your own before the meeting. Some of them are personal, and they are easier to answer honestly in your own head than out loud. *Discuss* prompts are for the room, and the good ones are built to split it. Having an outside facilitator might help ensure that the strongest voices in the room don't dominate.

- **A menu, not a checklist.** Pick three or four questions per session that fit your group. Trying to get through all of them is the surest way to have a shallow conversation about all of them.

- **It ends in work, not talk.** The final session is a working session, where each of you applies the framework to an outcome you actually own. That is where a book club turns into the thing the book is asking you to do.

You can run this in three sessions, grouped as below, or chapter by chapter if your group prefers seven shorter sittings. The three-session shape is the one I'd recommend for a busy team.

Session 1 - The Setup

Foreword and Chapters 1 through 3: the curve, the choice, and the river

These chapters establish the world before any decision gets made. The Foreword names the underlying problem (the gap between what your systems are supposed to do and what they actually do, and why speed turns a tolerable gap into a structural risk). Chapter 1 makes the case that the agentic AI capability curve is real and accelerating, and that the leaders pulling ahead are conservative institutions, not just digital natives. Chapter 2 says every enterprise is already making a retrofit-or-reimagine choice, deliberately or by default, and that neither end of the spectrum is safe. Chapter 3 reframes the choice as something you navigate continuously rather than select once.

Reflect (before the session)

- Two years ago, were you asking *when* about AI? What are you asking now? The book says the market's question has shifted from "when" to "why isn't this working yet." Has yours?
- The Foreword says "you cannot delegate this." Be honest about what you have, in practice, delegated. What AI decisions in your area are currently being made by your CIO, a vendor, an analyst firm, or a task force, that the book would say are actually yours?
- Chapter 2 argues retrofit failure is far more common than reimagine failure, but reimagine failure (Klarna) is the one that lives in executives' heads. Which fear has actually been driving your decisions: moving too fast, or moving too slow?

Discuss (in the room)

- The book claims most enterprises are "doing this wrong, in the same wrong way, for the same wrong reasons," and that the cost of inaction was negligible in 2024 but shows up on quarterly calls by 2027. Do you buy that timeline for *our* industry, or is it faster, slower, or wrong? Where does this room disagree?

- Pick a real, current AI initiative in our organization. Is it honestly a retrofit or a reimagine? The book warns that "you can spend a billion dollars on AI without reimagining anything." Would everyone in this room classify our initiative the same way?

- Chapter 3 says the leaders aren't ahead because they picked a better spot, but because they noticed faster when the spot moved. Do we have any mechanism today for noticing when our position has gone stale, or do we only reassess on the annual planning calendar?

- Dimon's line is "if you put your head in the sand, you will lose." The book's sharper question is whether your nearest competitor's next strategy deck still looks like yours. Name the competitor whose AI-related move would actually scare you. Are we watching them, or just watching the technology?

Anchor case. JPMorgan is the book's load-bearing example precisely because it is huge, regulated, and risk-averse, the opposite of a YOLO startup. Does that make the example more relevant to us, or easier to dismiss as "they have a $19.8 billion budget and we don't"? What is our honest version of "we are not Klarna"?

Session 2 - The Three Tensions

Chapters 4 through 6: workflow and workforce, context, and governance

Inside the boat, every piece of work is governed by three linked settings. Tension 1 (Chapter 4) is workflow and workforce, and the claim is that they are one decision most enterprises wrongly treat as two. Tension 2 (Chapter 5) is context, and the claim is that the scope of context you can set equals the scope of organizational reach you actually hold. Tension 3 (Chapter 6) is supervision, autonomy, and governance, and the claim is that you either build the rules into how the work runs or you bolt a lookout on afterward. The book's foundational argument is that these three move together, not one at a time.

Reflect (before the session)

- In your own area, are workflow and workforce one decision or two? Who actually owns the combined call, given that Ops, HR, and IT usually each own a piece on a different schedule?

- Chapter 5 says the hard part of context is making explicit what you have kept implicit in your own head for years. What is one thing you "just know" about how your function really works that you have never written down or said out loud?

- Chapter 6 says if your firm ends up bolt-on, it will be because the default carried you there. Which way is the default currently carrying your area?

Discuss (in the room)

- Chapter 4's bluntest claim: slotting agents into existing roles is "the slow path to irrelevance." Is that true everywhere, or are there functions in our organization where putting agents into existing seats is exactly the right and sufficient move? Where do we disagree?

- Chapter 4 names two shapes of reimagine: collapsing the org chart (Block) and rewiring the work inside the boxes (JPMorgan's KYC). If we reimagined one thing, which shape would it be, and what layer in *our* organization gets squeezed as a result? Can we say that honestly in this room?

- Chapter 5's uncomfortable claim: providing rich context to a team means giving up your exclusive hold on the knowledge your influence is built on, and that is uncomfortable in ways that have nothing to do with the technology. Is that discomfort real for people here? What context do you find yourself reluctant to make explicit, and why?

- Chapter 5's three levels of context: what you own, what sits inside the firm but outside your purview, and what sits outside the firm entirely. Which of our most important pieces of work actually require reaching into level two, and have we built the peer coalitions that takes? Or are we keeping work at level one because it is easier to control?

- Chapter 6 offers a tell: structural words (foundation, architecture, infrastructure) signal built-in governance; process words (council, framework, human in the loop) signal bolt-on. Listen to how *we* talk about AI governance. Which vocabulary do we use, and what does that reveal about which one we have actually built?

- The Air Canada, Microsoft Copilot, and Deloitte cases all failed not because the agent was too autonomous, but because "the thing watching the envelopes was the wrong shape." Where in our operation is the review machinery still built for the old shape of work while the work has quietly changed underneath it?

Anchor case. The book points out that the Big Four are selling agentic transformation while running it on themselves, so their case studies are "both descriptions and sales documents." When a vendor or advisor brings us their transformation story, how good are we at separating the description from the pitch?

Session 3 - The Practice and the Personal

Chapter 7 and the Epilogue: the plan you keep remaking, and you

Chapter 7 reframes everything: you are not managing AI, you are achieving outcomes you already own through a medium that now moves under you. The three Tensions collapse into one steering judgment (how far toward reimagine does this work need to go), outcomes get triaged by how fast the water is moving, and the plan is provisional by design. You sense at the speed of the river and commit at the speed your organization can actually absorb. The Epilogue puts the advisor's hat down and asks what all of this does to you, the person in the seat.

Reflect (before the session)

- Chapter 7 calls the snapshot mindset (deploy, then reassess on a schedule) "the one habit worth breaking on purpose." Be honest: is your current AI planning a snapshot wearing work clothes? When did you last re-answer the watching questions for an outcome you own?

- The Epilogue's central question: "Who am I, when the thing I was can be done by something other than me?" The author says peace with it is still a way off for him. Where are you with that question?

- The author admits he catches himself wanting to automate things simply because he can, treating human friction as a defect to engineer out. Do you notice that pull in yourself?

Discuss (in the room)

- Chapter 7's reframe: "you are not managing AI; you are achieving the outcomes you already own." Do we talk about AI as an initiative (task force, center of excellence, portfolio of pilots) or as a faster way to move the numbers we are measured on? Which mode are we actually in, and what would it take to switch?

- The book says the failure mode of large enterprises is not that they change too slowly, but that they *sense* too slowly. Do we agree that is our potential failure mode? Where are we reacting to a river that moved two channels over while we weren't looking?

- Chapter 7 insists the watch cannot be handed off to a market-intelligence team, a consultancy, an analyst subscription, or a research agent, because only the people steering an outcome can tell whether the water around it has changed. Are we already outsourcing our sensing, and to whom?

- The Epilogue argues that every decision about this technology is "a small moral choice in operational clothing":

which work goes to a machine, who you stand behind when a role dissolves, whether the people behind you inherit a ladder worth climbing. Is that the author overreaching, or is it the truest thing in the book? Where does the room split?

Bridge to the working session. The author says the single highest-value thing to do the week after finishing the book is to get the two or three executives who share your most important fast-water outcome into a room to read the water together. You are now in that room. Which shared outcome should this group take on first?

The Working Session

"Read the Water Together"

This is where the book club earns its keep. Run it as a working session, not a discussion. The goal is not consensus. It is to surface where this room genuinely disagrees about the water, because that disagreement is the most useful thing you will find.

Part A - on your own (before or at the start)

Each leader picks one outcome they personally own. Not "our AI strategy." A real number on your scorecard: cycle time, cost to serve, retention, risk reduction. For that outcome, answer:

1. **The water.** Is this outcome in fast water or slow water right now, and what's the evidence? Document-heavy

work, coordination across many systems, high-volume exception handling, and fragmented knowledge tend to sit in fast water. Work dominated by error costs, human trust, or the texture of a relationship tends to sit in slower water.

2. **The steering judgment.** How far toward reimagine does this work need to go this year?

3. **The three Tension settings that follow from that judgment.**

    - *Workflow and workforce:* agents into existing seats, or roles and work redrawn together? What gets squeezed?

    - *Context:* what does the team need to see that it can't today, and is that at level one, two, or three?

    - *Supervision, autonomy, governance:* how much runs without a human checking each step, and is the protection built into how the work runs or just inspected afterward?

4. **The help you need, and from whom.** From the CEO: air cover and the mandate to redesign rather than optimize. From the CIO or CTO: the platform and the built-in controls that let a team run wider without running reckless. From peer executives: the parts of the context that cross your boundary into theirs.

5. **The watch.** Which three questions will you keep re-answering for this outcome, and how often?
    - Has an AI-related capability crossed from demo to deployable against this actual work?
    - Has someone reset what "good" looks like for this outcome?
    - Has something you judged could wait started moving faster than you expected?

Part B - together (live)

Pick the one shared fast-water outcome that crosses two or three people in this room. Read the water together: where does it sit, how far toward reimagine does it go, where do the three Tension settings land, and what is the first move and who owns it? Spend most of your time on where you disagree about the water itself, not on the wording of the plan.

Before you close. This is not a one-time exercise. Put a date on the calendar to run it again, and remember the one rule that matters: the sensing has to outpace the moving. You watch at the speed of the river, and you commit at the speed the organization can absorb.

Finding the River

The Genesis of this Book

I wanted to write a business book for most of my career. The challenge was never subject matter.

I've spent decades inside organizations and have no shortage of things I want to say. My two real interests have always been corporate strategy and philosophy. Not philosophy as an academic pursuit, but the applied kind, the kind you reach for when you want to come at a practical point from an unexpected angle. I kept wanting to wrap a business argument in something with a bit more weight. For years, I could not find a compelling fit between topic and wrapper.

That instinct began a long time ago. In 1990, Ray Kurzweil published his first book, *The Age of Intelligent Machines*, a survey of where artificial intelligence had been and where it might be going, assembled partly from essays he solicited from contributors. He went on to become one of the best-known thinkers in the field. At the time, he was an inventor, placing a full-page ad in *Software Magazine*, inviting submissions. I was in my second year of university, taking a lot of philosophy and carrying a serious interest in computer science and AI. I decided to submit something that I had originally written for a Mind and Consciousness course, a Socratic dialogue between a 'human' computer (AI) and a materialist philosopher. It contained a fairly technical argument about the future of intelligent machines, dressed up as a philosophical conversation. Almost a year later, Ray sent me two copies of the book and a check for seven hundred dollars (a lot of money to a university student at the time) and said thank you.

For the next three decades, I never found the business equivalent of that wrapper. Then, not long ago, it arrived. The thing I kept circling was that the agentic AI landscape is moving so fast that a strategy you commit to today can be stale before you have finished putting it in place. A current you cannot stand still in. A river that does not wait. Once that image was in my head, the rest of the book started to quickly flow from it (pun not intended).

So the conception was mine, and it was mine well before I had any help executing it. But I did have help.

The same principles and practices for managing the current of agentic AI that I detail in this book were used extensively in its making. The largest and hardest share of that work was research and verification. I ran background research across all three of the major models, Claude, ChatGPT, and Gemini, specifically so I would never be trapped inside any one system's blind spots. What I found is that they rarely fully agreed with each other. Usually this was not because one of them was inventing facts. It was because they had different access to information, drew on different sources, and weighted them differently. I was really interested to discover how true this is of human sources as well. I would often read two published accounts of the same event and get two different versions, both of which could not possibly be correct. Because of that, I spent at least as long verifying as I did researching.

Related to that, I worked hard to get the factual record right as of the middle of 2026. I will correct anything that turns out to be wrong in future editions, with apologies in advance if it comes to that. The cases and figures are here just to illustrate the arguments made and none of the core ideas stand or fall on any single one.

I also used AI as a collaborator in the writing process. The voice is mine, and a great deal of the text is literally mine, but there are passages where the agent led. Here is why that doesn't trouble me. The fluency was never the part I worried about getting right. There are many people who write more beautiful prose than I do, and the machine is better at the surface of it than I am too. What a model cannot do is decide what is worth saying to a working executive, hold an opinionated point of view together across a whole book, or know from experience which of its own fluent suggestions is subtly wrong.

And the machine was nowhere near the last word. Once the manuscript existed in draft, it went to more than a dozen human readers, people whose judgment I trust, who came back with numerous corrections, objections, and changes. I read every one of them carefully and made the final edits myself.

There are things I simply did not hand over. I did not delegate the thesis. I never asked a model what this book should argue. I did not let one choose which examples mattered, which claims survived, or where the line fell between quality arguments and needless filler (the killer of so many business books). When the machine handed me a sharper sentence, I took it. When it handed me a plausible claim I could not stand behind - no matter how intelligent it sounded - I cut it. When it told me how great my idea was, I largely ignored it. When it criticized me - which I specifically instructed it to do frequently - I paid attention. Using agentic AI in this fashion is what I have spent this whole book arguing that leaders must do to navigate these systems. Workflow. Context. Governance. You can hand off some of the capability. You cannot hand off the accountability.

This book came out of my mind, based on my experience, expresses my strong beliefs about what it will take to be successful in this new era, and truly comes from a heartfelt place. I offer it to you in that spirit. Onward!

[For those interested in a more tactical/technical description of how I used AI to assist writing this book, please reach out to me at blaine@riverdoesntwait.com.]

Acknowledgements

This book could not have been written without the lifelong support of my wife Wendy Mathieu, including the more proximate support she gave me in enabling me to spend time on this book, during a career break, as opposed to immediately engaging in my next corporate adventure.

As for the early development of my mindset toward business and my aptitude for entrepreneurship, I have to give most of the credit to my father Guy Mathieu and my father-in-law Ernie Daviduck (both successful businesspeople), and to my long-time mentors Leo Houle and Randy Eisenbach, corporate executives who each contributed in their own very significant ways to the person I am today.

I also need to credit my son and content guru, Aiden Mathieu, who provided the encouragement to use the river/boat concept as the foundational analogy for the book.

My colleagues at the Post-Industrial Institute and the Mindful AI Foundation were definitely central to my thesis formation around the topic of the transformation of organizations via agentic

AI. I want to particularly call out Frode Odegaard and Rao Mikkilineni as especially impactful.

Also thanks to my design partner over the last 15 years, Diana Rudolph of Growth Catalyst, for her assistance with the cover art and layout. See her work at https://growthcatalyst.us/

Finally, to the hundreds of people I've worked with and for over the years. For your leadership, direction, advice, and friendship, I just have one thing left to say: Thank You.

About the Author

Blaine Mathieu is a former Gartner analyst and multi-time CEO, Chief Marketing Officer, and Chief Product Officer at both public tech giants and private enterprise software and Industrial-focused startups. His clients have included the likes of ABB, France's Total, Ford, China State Grid, SoftBank Corp, and many others. Blaine's work has spanned North America, Europe, and Asia, and his lifelong passion for corporate strategy, AI, and software generally makes him a sought-after speaker at global industry events. Based in the San Francisco Bay Area, Blaine's first work in the field of AI was published in Ray Kurzweil's 1990 book, *The Age of Intelligent Machines*, and he has been an avid follower of the space ever since.

On a voluntary or advisory basis, as of the time of this writing, Blaine was associated with the following organizations:

Post-Industrial Institute: Based in Silicon Valley, the Post-Industrial Institute is an independent research institute. Its mission is to develop a post-industrial management science and tools. It puts research into practice by helping leaders build the organizations of the future. Blaine is the Host and Moderator of the Post-Industrial

Summit event series as well as the Post-Industrial Voices podcast. https://post-industrial.institute/

Mindful AI Foundation: The Foundation is a call for a pivot in our quest for human-level intelligence. Human-level intelligence is not only the power to speak, predict, or imitate. It is the capacity to remember, to govern, and to remain worthy of trust. Its vision is to help build the constitutional infrastructure that makes that standard possible - so that as intelligence scales, it does not merely expand in power, but deepens in responsibility. Blaine is a Fellow at the Mindful AI Foundation. https://mindfulai.foundation/

Human Intelligence Institute: The Institute proclaims, protects and preserves the value and meaning of human creativity in the age of AI and provides toolsets to help Human Creators do the same. Blaine is on the Advisory Board of Human Intelligence. https://humancreator.org/

Creatorbase: A community and media organization for creative businesspeople and technology industry leaders everywhere. People take action to own their business domain expertise and help others succeed with Creatorbase®. Blaine is a Founding Member. https://www.creatorbase.com/

Reach out to Blaine at blaine@riverdoesntwait.com and you can always find the latest about him on this book's website:

https://riverdoesntwait.com/

www.ingramcontent.com/pod-product-compliance
Lightning Source LLC
LaVergne TN
LVHW100311210926
848480LV00003B/1099